Charles Patton

Visionaire

The Evolution of Technology
from the 1960s to Today

Charles Patton: Visionaire – a tale of persistence, resilience, and creativity

Notice

To protect individual privacy, some names and identifying details have been changed. People's names have been included when sharing or giving credit, while criticism is masked by omitting or disguising names. This memoir aims to present truthful recollections and opinions without intending to offend anyone or cause embarrassment. Also, in describing my harrowing experiences, please understand that I have not exaggerated and, if anything, understated them. There is no exaggeration in any of this account of my work life. Despite efforts to be factual and honest, errors and omissions are inevitable. I apologize for any inaccuracies or misremembered details. I encourage readers to point out significant flaws so that they can be corrected in future editions. I apologize to anyone deserving recognition who may have been inadvertently omitted due to my aging mind's forgetfulness, as it was never intended to disregard anyone.

Short Mystery Press
Written for and owned by:
Applied Market Solutions, LLC
Charlespattonbooks.com

Book cover by Book Design Company
Book template by usedtotech.com.

Role of Artificial Intelligence (AI) in this book: I found openAI's chatGPT™ and Google's Bard™ helpful in cleaning up, simplifying, and clarifying my writing and in finding or checking sources. The owner of this manuscript is responsible for its contents.

I coined the term "Visionaire" to encapsulate my career as a visionary of extraordinary projects. This new word reflects my ambition to serve as a beacon of light and inspiration, embodying a career path marked by bold foresight and transformative ideas.

CONTENTS

BEGINNING MY JOURNEY

This memoir is about the early days of computing, my involvement in its evolution, and how I grew to lead large projects for branded companies. It also reveals how I grew from a humble programmer to a leader of Information Technology teams for world-known Travel businesses. And it includes a touch of career advice along the way.

I have seen and had to adapt to remarkable changes in technology over my lifetime.

Growing Up

I was born in 1942 before there was TV, portable phones, and air conditioning. Growing up as the second of six children, I was fortunate to have educated and industrious parents who had varying viewpoints. My father was a Republican and a chemical engineer, while my mother was a Democrat, housewife and entrepreneur. Their differing perspectives fostered lively discussions and exposed me to diverse people, facts, and opinions from an early age.

Moreover, I spent most of my childhood in rural areas, where I honed a strong work ethic working on farms and in a variety of part-time jobs. These experiences instilled in me the value of hard work, perseverance, and adaptability, traits that have served me well throughout my life.

In high school, I worked on a farm in the summer and sometimes on weekends. The farmer I worked for was the most fantastic person I have ever met. He could do anything.

on his farm. He worked seven days per week running his farm. He welded, cared for his animals, kept accounting books, butchered meat, repaired engines, and excelled at every skill you can imagine. Then, a year before he retired, he showed me the 10 acres next door and told me he would build his retirement home there. Not only did that one man build that house while running his farm, but he also made every piece of furniture– tables, chairs, dressers, beds, etc. When I saw what he had done, I decided I wanted to be like him, where I would learn to do everything possible. My goal was to be the Jack of all Trades in every job and my personal life, including learning how to build furniture.

I grew up in the "BM" era, meaning "Before McDonald's" -- pre-fast food. It was when the best source of facts was the Encyclopedia Britannica or other books, well before "fake news." The nutrients in our Midwest cropland were not yet exhausted. Round-up™ had not been created yet. All food was naturally organic. All meats and eggs were naturally free-range, and all fish were wild caught, often by me.

At the same time, our water came through the house in lead pipes. I often played with a half-liter of mercury that my chemist father gave me, with no noticeable side effects (yet). And I had my first gun at age 12 and never once thought of shooting another person, even when bullied mercilessly or losing in childhood fistfights.

I received my education by attending seven different grade and high schools and four colleges, mostly public but four parochial, across nine cities in three states over seventeen years. The quality of my education varied but was mostly good, with nuns giving me my best foundation, including biology, trigonometry proofs, Latin, and how to dance.

I had my first taste of organization in the Boy Scouts, where I became a Patrol Leader, a role like a Sergeant in the Army, supervising a small group of other Scouts while working toward achieving the level of Life Scout, one step below Eagle Scout. I still regret that I passed all the requirements for Eagle Scout but aged out before the award could be approved. However, I took my leadership role seriously and took the principles of scouting even more seriously. I have followed them throughout my life: "A Scout is trustworthy, loyal, helpful, friendly, courteous, kind, obedient, cheerful, thrifty, brave, clean, and reverent."[16] I also immensely enjoyed camping, the smell of campfires, roasted marshmallows, hot chocolate, and listening to ghost stories. And they taught me to dig a shallow trench around my pup tent before retiring – it protected me more than once from seepage on rainy nights – and demonstrated their motto: Be Prepared!

College

As you will read, I went to college and barely made it out, paying the cost myself through jobs, loans, small Pell grants, what little help my parents could provide, and my final semester courtesy of my future wife's remaining bank balance. Because my direction was unclear after high school in Arlington Heights, IL, my mother pressured my father to escort me to Navy Pier in downtown Chicago and enroll me in the University of Illinois. She did not want me to be drafted or enlist, as I was entertaining joining the Navy or Illinois National Guard. I had already received a letter from the National Guard to report for training. The Draft Board exempted College students from military service if they were actively enrolled full time. Until then, other than taking pre-college classes and the PSAT, SAT, and ACT. I had not seriously considered college as an option for me.

Fortunately, University of Illinois' tuition was much cheaper than today. It was $125 per semester, equivalent to $1,250 in 2023 dollars, compared to their 2023 base tuition of $6,237 per semester.

I saw myself as a "common" or "regular" person. I never fit in with the billionaires I later came to know. I did not fit in with executive management until I entered that zone, and then only fit so-so. I worked my way up from the bottom and reached a respectable level by the end of my career, given where I started and how naïve I was when young. Along the way, similarly "common" people often helped me.

Early Learnings

Because of my simple origins, I lacked insight into the possibilities "out there." I viewed myself as a team player, a subordinate, a partner, a sidekick, a collaborator, or a co-worker. I liked football because if I missed a tackle, one of my fellow players would get him. I liked the idea of strength in numbers. It took a long time before I emerged as a leader, and even then, I was more of a circumstantial leader, stepping up when needed or someone asked me to take on a large project because I had demonstrated the organizational skills.

What I learned along the way, and wish I had known earlier, was that you will never get into a leadership role if you cannot see yourself there. So, you must know where you want to go and commit to that for it to happen. To achieve success, there are four key steps. First, you must have an awareness of the possibilities that exist, understanding the different paths you can take. I missed this step until much later in life. Second, make a commitment to a specific objective or one of the possibilities you envision, providing focus and direction. Third, work diligently to gain the experience and skills needed to support your goals. This requires dedication and effort. Lastly, apply the experience you have gained to seize significant opportunities when they arise. By following this process of awareness, commitment, hard work, and application, you increase your chances of achieving success and reaching your desired objective. Good Luck is when Preparation meets Opportunity, a saying attributed to the ancient Roman philosopher Seneca the Younger.

As you will read, my first career job was as a programmer trainee. I learned early to love programming and appreciate other programmers. But I also learned to appreciate all the others who help computer systems come to life -- planners, trainers, procedure writers, testers, computer operators, and all the others who have worked for decades to get society to the point where it is technologically today. Programmers are often recognized for their intelligence, logical thinking, and ability to organize and learn about a wide range of topics.

However, many intelligent, analytical, and organized people are not cut out to learn to program. It is a unique skill, just as not everyone is cut out to be a successful musician.

I very much respect, among this vast group of technologists, the innovators who came up with the core ideas. These are very often unsung heroes. These "mini-innovators" are the ones who turn a good product into a superior one. Look at any marvelous piece of technology, and you will see hundreds of innovations whose creators you know nothing about. For example, who first thought of Wi-Fi? Vic Hayes gets credit because he chaired the IEEE committee that created the 802.11 standards in 1997.[24] What about the other people on that committee? Or was it Hedy Lamarr, the actress, who was awarded U.S. Patent No. 2,292,387 in August of 1942 for a frequency hopping communications device to protect guidance systems on torpedoes, which may have led to the development of Wi-Fi, GPS, and Blue Tooth?[6] Or, was it all the thousands of people who took what Hayes and Lamarr started and transformed those ideas into what we know today as Wi-Fi? Consider all the historical work leading up to the iPhone. My career has been a mixture of innovations, a few "firsts," the occasional initial big idea, but more often, mainly "mini-innovations" along the way.

Charles D. Patton

INTRODUCTION

Life is full of those pivotal moments, such as where to go to school, whether to join the military, whom to marry and, of course, what career to pursue. Often, such significant decisions are driven by what others decide, by coincidences, accidents, fate, or other less than clear-cut conscious decisions. This was certainly the case for me.

No matter how it happens, your career choice will determine where you live, the kind of work you do, whether you are happy working, your growth potential over the years in that career, and eventually the quality of your retirement. Monumental, right?

And the pivotal decisions that determine your career mostly happen when you are young unless you phase out of one career into another (more on that later). To end up somewhere in any career, we all start on a "path." Some people find their path early, others late, and a few never. I found mine early, but my path was overgrown and weedy. I was not sure of my path at the start or confident I could follow it to the end. So, my intentions evolved along the way as did my skills.

Book Credits

No one creates a book by themselves, and this one is no exception. I would like to recognize my wife, who has been my partner throughout my career and has put up with me for 75% of my life. She reads what I write and makes many improvements. Also, I want to thank my family for their ongoing support. Lastly, I want to thank the programmers who wrote my writing tools, including Alex Shevchenko, Max Lytvyn, and Dmytro Lider, who founded Grammarly, the programmers of chatGPT, who made finding sources easier, and Writers Workbench's creator Greg Oij, who had nothing to do with this book, but who has followed a technology career path like mine.

When I pressed chatGPT on who created their AI tool, it responded: "While it is difficult to attribute the creation of ChatGPT to any specific individual, there are several key individuals associated with the development of the underlying GPT technology, including but not limited to researchers such as Alec Radford, Ilya Sutskever, and Wojciech Zaremba. However, it is important to acknowledge that ChatGPT is a product of collective work and the expertise of many talented individuals at OpenAI." I find it interesting that even an AI tool recognizes the broad number of people involved in its creation.

Why This Memoir?

As I said above, this book is about the early days of computing, how I got into it, where it took me, my unique part in the industry, and where I ended up. It is also about creating, organizing around, and implementing large projects in rapidly growing industries, and about thinking big and taking risks. I will share a few management skills I learned along the way. Finally, it is also about how others helped me grow in my career and what was accomplished, not by me alone but by all those who gathered around me to achieve remarkable things. So, thanks to everyone who gave me a job along my path and all those who shared my journey.

Hordes of people were working in all aspects of computers before me, and even more came into the industry after me. Many collaborated with me during my contributions over the years. I never knew the big "financial winners," like Bill Gates and Steve Jobs, who became billionaires. I knew and worked for a handful of businesspeople who benefited from my efforts and became billionaires, but not me. My main point is that the vast majority of the people I knew and worked with contributed extensively to the growth of the computer industry, yet they never came close to becoming wealthy. I have done okay and, like most other contributors, made a decent living, but only a few hit the big rewards.

I have remembered as many names as possible of those who helped me throughout my career or who worked with me on various projects. Sadly, in too many cases, the names of important contributors have fallen off the back of the truck that is my mind. For those I complimented, I included their full name when mentioning them. For those I criticize, I only use their first name and the first initial of their last name or omit their name altogether. My mother always told me, "If you do not have something nice to say, do not say anything at all." I tried to follow her advice.

It Takes All Kinds

6.5 million new people are still entering the computer industry every year, but the industry is vastly different from when I started. And it will not be the same tomorrow as it is today. Truly little is the same now as when I started. Few industries have changed more rapidly than IT and for such a long time. Physically, early computers filled entire rooms, required special clean and chilled environments. Now they fit in your palm.

Some of my predecessors were hardware specialists; others took the software route. Some Software programmers programmed Operating Systems, like Bill Gates and me (once), while others wrote Applications, like me. The first Applications were mostly basic: mailing lists, payroll, tabulating, and the automation of other manual processes.

Later came complex systems needing multi-processors, vast storage devices, and more. Then came the internet and then cell phones, which exploded applications to a degree unimagined even by most forward-thinking of the early computer people. Now those apps are spreading across the world. Satellites are now encircling the globe overhead, soon giving internet access everywhere. It is sad and unconscionable that with all this knowledge available anywhere in the world we still have people starting wars, living homeless on the streets or in hovels, and unable to obtain basic health care. This book is meant to chronicle the vastness of the changes in computing and business over my lifetime.

My Inspiration!

The very first step on my career path began in a high school AP physics class in my Senior year. My GPA was middle of the road, but I did have a talent for being what some call "test-wise." So, I managed to crush my PSAT, SAT and ACT, scoring in the 96th percentile in math and the 88th percentile in English. Those scores qualified me for that class, my one and only AP class, and would later barely get me into a state college. My GPA in high school was 3.2 on a 5.0 scale.

My career inspiration came when my Physics teacher showed the class a film about programming a massive computer, an RCA (model 501), for the military with MIT's help. I had never heard of a computer until then, but I immediately grasped its potential. I recall one scene in the film of a massive wall of electronics and men working on ladders. The size and power impressed me, and I began to imagine all kinds of potential uses for this new device called a Digital Computer.[20]

This class also introduced me to the characteristics of waves, which would benefit me later when I took a class on Analog computing and, even later yet, when I later experimented with IBM's Quantum computer.

In my math classes, where I sat in the back row by the window, I daydreamed about how computers had so many possible applications. Sometimes I was so distracted that Dr. Shirley, my math teacher, would throw chalk at me to refocus my attention on his lessons. One application I was imagining was the ability to simulate or electronically play a football game -- design plays, direct the players, and run the whole game yourself. I was an ex-football player who still loved the game and saw how the game's logic and strategy could be captured in programming and run by a computer. I wished I had pursued this idea, but I did not. In 1988, EA Sports preempted my hope of pursuing that possibility with its John Madden game, which still pulls down a cool $600 million per year. Although it still lacks features that I envisioned.

I was a giant fan of Arthur Conan Doyle's Sherlock Holmes mysteries. Sherlock often spoke about decoding secret messages or ciphers. I could envision, in 1959, a computer that could create unbreakable encryption of secret messages. This I did pursue (more on this later). When I first thought about it, I was too young and inexperienced to pursue my ideas about encryption. However, I was interested in electronics, which has similarities with programming, and I delved into that early.

A TV repair shop was near my high school, in a small strip center, next to a small diner where I went occasionally for lunch. I had built a radio in middle school, so I had a rudimentary knowledge of electronics and the internal parts of radios and TVs. Envisioning a connection between the insides of TVs and Radios and the insides of computers, I walked into the shop one day. The guy inside was amiable and showed me what he did. He then offered me a part-time job because he was loaded with repair jobs. I accepted and started an unofficial apprenticeship in TV repair and electronic circuits.

I would drive his panel truck to pick up TVs needing repair and return them to customers after they were fixed. I had a uniform shirt with the company name on the back, but the name embroidered above the front pocket was Steve. I presumed that Steve was my predecessor. I never got used to people calling me Steve.

In addition, he taught me how to repair not only the easy-to-replace vacuum tubes but also how to solder and replace resistors and capacitors in televisions. He taught me to be safe and avoid touching the Power Supply because even if the TV was unplugged or turned off, it still held high voltage. A few times, he sent me out to do repairs at people's houses. My boss told me to bring the TV back to the shop if it was beyond my skill set, but usually, I could fix them. I only needed to bring one in, and even in the shop, I could not fix it. My boss repaired it. It had a bad capacitor which is hard to diagnose, and a repair that required soldering, which I knew how to do but was not something I should do in someone's house.

Undiagnosed Dyslexia

I struggled in math, even though I had a keen interest in it, due to my having dyslexia and not knowing I did. Dyslexia was discovered in 1881 by Rudolf Berlin, a German ophthalmologist, and professor in Stuttgart. Still, I did not know I had it or anything about it nor did anyone around me.[14] At the same time, this malady, which made it hard for me to organize my thoughts around any subject, also impacted my English grammar and writing performance, except, oddly, on tests like the SAT. My weaknesses in reading and writing hindered me in all my classes. It is a wonder that I graduated from high school, let alone college. The following example exemplifies how bad this handicap was.

In my sophomore year at the University of Illinois, I was required to take two semesters of Rhetoric, in which I performed miserably. Then, at the end of that year, I was required to pass a proficiency test, which I failed. The school then mandated that I take a class called "Remedial Rhetoric," which I did. I was assessed again, and again I failed. Finally, as the University's last attempt to move me along, I was required to take another course called "Advanced Remedial Rhetoric." Even with my poor rhetorical skills, I understood the irony in the title of that class. That class directly targeted the test -- what one needed to know to pass the test, which I finally did. Even so, I still place commas in the wrong places.

Decades later, I learned that I was dyslexic, and it became apparent that my struggles in school could be attributed to this condition along with me being the youngest in my classes throughout grade school and high school. I struggled to read and memorize anything and was saved along the way only by being "test-wise." Over the years, I grew to understand that I learned best when doing something manually rather than trying to absorb it from a book. Despite my handicap, I did learn to speed-read IBM manuals later out of sheer necessity. Today, when I read, I still miss certain meanings. And, sometimes, I still put order in the wrong words (kidding!).

As an example of my being "test-wise," I had taken German for two years in high school, learning not one word or anything about German grammar. One year of a language in high school is treated as one semester in college. During my initial admission into the University, they insisted I take a German Placement test to see where I would fit in their four-semester language requirement. I offered to change languages, but they would not consider that until I had taken the German proficiency test. When I sat for the test, I found that even the instructions were in German, and I understood none of them, let alone the questions, I still passed with excellence. Weird!

As a first-year student, they started me in third-semester German. I got a grade of D in the third semester, and before the fourth semester, the instructor, who would teach the fourth semester, saw my hopelessness, and kindly made me an offer. He told me that if I attended every class and took every test, he would pass me with a D so that I would complete the requirement. This is an excellent example of how I struggled through school. It is also the first of many examples of an angel who helped me along my path, for otherwise, I might have flunked out, been drafted, and potentially killed in Vietnam.

Transfer to Champaign Illinois (and ROTC)

By 1962, I had finished two years at the University of Illinois' Navy Pier campus in Chicago. I transferred to Champaign with fifty-seven total hours (three credit hours behind expectations for students halfway through) and a GPA of 3.017 after just getting off probation (probation started at 3.00). So, I was in danger of flunking out.

The university rule stated that if I had fewer than sixty credit hours as a junior, I would be required to enroll in Reserve Officers' Training Corps (ROTC), which involved undergoing military training for four semesters on the campus in Champaign. This program allowed me to select one of the five Armed Services for my training.

Upon graduation, I would be inducted into that service and likely sent to Vietnam as a second Lieutenant. The mortality rate for second "Louies," as they were called, was said to be on the high side because they were often on the front lines. For as long as I was enrolled as a college student during the early 1960s, I was exempt from military service until graduation. However, I was not exempt from ROTC, and I had to petition an Assistant Dean to be relieved of the requirement. This involved being interviewed and presenting my case for exemption.

I went to that office, and a young man behind the receptionist's desk handed me an essay form and told me to write why I felt I should be exempted. As I sat writing up my reasoning, another student who told me that he had 59 hours went into the dean's office to be interviewed. He came out again in ten minutes, grumbling. He said he did not get excused. I saw what little hope I had fading away.

I related my fears to the receptionist. He asked me to show him my essay. After reading it, he said it would not do, tore it up, and handed me a new form. He told me what to write to describe how I needed the time to finish my core requirements, which was true. It worked. The dean let me off. I was exempted only because that young man intervened to help me. He was another angel who helped me along my path. And once more, I was saved from the clutches of Vietnam.

College Dropout

After my first semester of my third year in college, I had to drop out to earn money. A friend was working at the Illinois State Bank and told me they were hiring. I worked there for a semester. They were very much into training, and they taught me how to operate the bank's Burroughs ledger accounting machines, the coin counter, and the Addressograph, as well as how to be a teller, my main job. These machines did tasks that computers would soon take over, but they also taught me certain business processes, like maintaining a mailing list, the difference between a debit and a credit, and ledgering.

While working for the bank, besides taking American Banking Association training classes, I took two evening classes at the Illinois Institute of Technology (IIT) to keep up my college studies. One course was in Analog Computing, and the other was in Machine Tool Programming – both of which interested me. Being enrolled even for two classes kept me exempt from military duty.

Although it was a common strategy among young men at that time, I was not actively dodging the draft. My decision to stay in college stemmed from the belief that the military would not allow me to make the most of my talents. While I anticipated serving once, I completed college, I also held the hope that the war would be over by then. By not serving in the military, which would have taken two to three years of my life, I could get a head start on gaining early career experience.

The analog computing class took place in a computer operations room that housed a big digital computer on the Illinois Institute of Technology's campus south of the loop in Chicago. I was fascinated by the big computer because, when its access doors were open, I could see two-foot-tall glowing vacuum tubes. They were part of the power supply. They looked like giant versions of the vacuum tubes found inside old radios, with these having glowing filaments emitting an eerie green glow.

The analog computer used a plug-in board with holes into which we would plug resistors, capacitors, and connecting wires. The difference from a digital computer was that the changes in the fluctuations of the voltage and amperage made the computations. We observed the results on an oscilloscope, represented as a sine wave.

Addition would be achieved by combining two waves to make a higher wave and subtraction by inserting a resistor to reduce the wave's height. It worked but struck me as limited in its application to common problems. Analog computing lost its race with digital computing over the next few years due to digital computing's greater efficiency. Analog computers had limited output precision due to component accuracy, no memory, were power-hungry, and were designed for single tasks. In contrast, digital computers were easily programmable, precise, and could store programs and data.[27]

I related to the analog form of programming better than the other students in the class because I worked as a TV repair person in high school and learned about resistors, capacitors, and vacuum tubes. Note: The latest Quantum computers may bring back analog computing.

The other class taught me how to program a seven-axis machine tool. This "tool" filled a whole room and had robotic arms. The machine tool could move its arms with smaller tools attached to them in seven different motions – The 7-axes meant the machine arms moved right-left, top-down, and back-front, followed by the axes defined by the spinning of the tool, rotation of the part, rotation of the tool head, and movement for clamping, reclamping, or removing the part.[21]

This device could readily turn a cube of solid steel into a car engine block with all its holes, openings, and curves and do it in a couple of hours or less. For example, when a machine tool program told the machine to drill a hole in a block of steel, its arm would reach out to the appropriate drill bit among the many tools hanging on the surrounding wall, attach the bit to itself, then move over to the block of steel and drill the precisely-sized hole in the exact location to the right depth as it was programmed to drill. The massive machine tool could do all the work of a skilled machinist but faster, more accurately, and repetitively. I might have followed this as a career, for it became huge in the automobile manufacturing industry, but I did not. Instead, I dreamt of digital computers, football games, and encryption.

Marines

At one point, before I met my wife, my funds were too low to sign up for my next semester. I went to see the College Loan officer; he turned me down for another loan. I found a brochure promoting the Marines on the way out of his office. They had a program where they would pay for my remaining college if I joined up. I would attend basic training during the summer and be promoted to Second Lieutenant upon graduation. I seriously considered this. I filled out the form and, with the pen in my hand, was about to sign it but hesitated. As I reflected, the thought came into my mind to make one more run at the loan officer before I signed. I returned to the loan office and told the officer what I was about to do. He said, "Do not do that! Okay, I will give you another loan." Again, I was saved by a caring person from a fate that included the hazards of Vietnam.

Choosing a Major

When I went to register for my junior year, I still needed to figure out where my career was headed. At the registration desk, they asked me what my major would be. I was flabbergasted and felt that I had to decide what the rest of my life would be -- at that very moment. So, I asked what my choices were. The registrar pointed to a placard that listed about forty majors in the College of Liberal Arts & Sciences.

Picking up the card, I excused myself from the line and began walking around in a big circle studying the card. I could not find any subject in which I wanted to major except Math, and I had no confidence in my ability to succeed in that major. After about fifteen minutes, I choose Psychology because it sounded easy. But, wow, that supposition turned out to be very wrong. Psychology entailed much reading and memorizing, which was a lost cause for my dyslexic brain – an issue I still had not recognized back then.

Even though I liked math, my grades in that subject were poor. In my first semester, I took a five-hour class in Analytical Geometry and failed it. For the next semester, I thought, if I take it again, I should do better and average my grade up. I took it again in my second semester and got a D. So, at that point in math, I had ten course hours of D minus.

There were no Computer Science curriculums in universities yet, and barely any classes in programming (Pascal, LISP, and Fortran). Hence, the career paths to computers were either through pursuing a Math or Physics degree, with the latter focused more on hardware. Most schools did not yet envision computers as a career, an industry, or the basis for a curriculum, let alone a degree major. So, it is understandable in hindsight that I did not know yet that programming would capture my interest and was unsure what my career path should be. My major was psychology and I thought math might survive as my minor.

Marriage and First Child

Around this time, 1964, I married my girlfriend, a beautiful coed and ex-high school cheerleader. Until that point, I had led a highly active social life, frequently indulging in beer consumption and intermural sports. Looking back, I regret the excessive drinking and wish I had made different choices.

It was then that the Draft Board removed the exemption for college students, so being married gave me a new deferment from military service. By the end of the year, my wife, whom I had met on campus, and I welcomed our first baby—a boy named Christopher. Having had a child again extended by exemption from military duty, just as the Draft Board removed the exemption for married men without children.

The responsibility of being married and becoming a parent was the catalyst that caused me to grow up and become serious about life and studies. On top of these challenges, my bank account was empty, and my options for borrowing more money from the school were exhausted. My wife gave me her last money, $180, so I could enroll for my final semester while she moved in with her sister in Chicago. Having family responsibilities, running out of money, and being on the verge of failing to graduate, was a sobering combination of events and a strong source of motivation. I was always motivated by desperation.

.

CHAPTER 1
MY FIRST STEPS INTO I.T.

As mentioned above, I met my wife in college and married in 1964. This was in the Spring of my senior year. I took four and a half years to graduate because I had to drop out for that one semester to earn money. So, my senior year was two semesters separated by a summer break. During that summer, my wife, who had been living with her sister and brother-in-law, and I moved into a three-room, third-floor walk-up apartment on the north side of Chicago. My wife had been working for Union Carbide as a Secretary, but we had truly little money, so it was essential that I get a job to make ends meet for that summer. We had already moved into our tiny apartment, but to make next month's rent, I needed a job. I spent the first week of our summer break job-hunting. I was willing to take any job. I walked the streets of the loop looking for a job for an entire week, applying to dozens of companies without any of them showing the slightest interest. I was worried.

By 3:00 PM on Friday, I was jobless and massively dejected. However, considering the precariousness of my family's situation, I mustered the strength to try one more company. The building in front of me was the Monadnock building, an impressive 16-story building on the south end of the loop. It is unique because it is the city's last load-bearing" structure, meaning that the massive stone blocks that form its exterior carry its load.

I walked in and looked at the sign showing the tenants. I saw one on the fourth floor called Statistical Tabulating

Company (Stat Tab). I had taken a psychology class in statistics, so I thought it might be a good prospect for me. I stepped into the elevator and, since they had operators at that time, I requested the Statistical Tabulating department.

The operator stopped on the third floor and motioned me out. I was puzzled because I thought I saw it was on the fourth floor. The elevator left me standing alone. I looked to the left and right in an empty hallway. All I saw were solid office doors with only numbers and no names, titles, or companies. I picked the one in front of me and knocked. A man inside told me to come in and asked me what I wanted. I told him that I needed a job badly and hoped my class in statistics might qualify me for something with them.

After listening patiently to my request for a job, he told me to go down the hall to the door at the end and tell them that the Manager of Operations sent me to be hired. I started walking while he picked up the phone. They hired me for the summer. And the pay was more than I had hoped for, $350 per month, enough to get by. I was on cloud nine and floated home to tell my wife. We celebrated by me walking several blocks down the street to buy a television for $20 that I saw advertised on a 3x5 card at the grocery store. I did not realize how heavy those old TVs were until I had to carry one for three blocks to get it home. I almost collapsed but made it.

To keep up with my studies, I took a correspondence course from the University of Illinois on Comparative Economic Systems.

My First Position in Data Processing

My very first job at Stat Tab was survey coding. I was one of six clerks who would go through paper surveys and mark the open-ended written answers according to a checklist we were given. We were turning written answers into single-digit codes. For example, if a person answered a question that asked what color shoes they wear, if they wrote Brown, we would mark on the survey the number 1. If they wrote Black, we would write the number 2, and so on. Numbers were easier for tabulating machines to process than words, especially when opened-ended answers were sentences, not single words.

The surveys would then go to keypunch operators, who would punch holes into "IBM Cards," corresponding to the codes we wrote. After the cards were punched, another set of operators would "key verify" the cards by repeating all the same keystrokes of the keypunch operators without punching any holes. The verification process typically found 2-3% of the punches in error. The erroneous pages and cards would then be returned to be keypunched again and reverified. The devices used were the IBM models 028 and 032 keypunches, and the IBM model 051 Key Verifier. These were the workhorses of the early data processing days, with dozens of machines in use simultaneously – raising a din of clicking and clacking. The IBM 029 looked like:

Each card held eighty columns where a punch out could designate a number or a letter in that column. For instance, in this system, numbers such as 1, 2, or 3 were represented by punching corresponding holes in the first, second, or third positions of a column. Similarly, a combination of two holes in a single column would signify a specific letter of the alphabet.

The complete set of punch combinations representing numbers, letters, and special characters is called Hollerith, named after its inventor. Hollerith was the system everyone used for coding data into punched cards. This encoding structure is still included in programming today.

Once all the cards were punched perfectly, they might be sorted by a card sorting machine to put them in a particular order, and then would be run through an analog device called an IBM Model 407 wired accounting machine, and the results printed out on paper or check stock. This device used a fixed typeface known as E13B. This typeface was specifically designed for magnetic ink character recognition (MICR) designed particularly for checks and other financial documents. At times we had to design check stock using a special 132 character-long ruler to make sure the data would line up correctly. The 407 looked like this:

This machine was programmed using plug-in wires (like old switchboards used), which bridged electrical current from where the machine read a punch to "accumulators." The device also translated the punches into their corresponding characters so that words could be printed. Entire reports, like Profit & Loss Statements, could be printed based on these plug boards. The machine could also print out all the checks for a payroll. This was an early form of programming without a programming language per se.

The printing mechanism was like a typewriter. The printer used paper for reports that was 132 characters wide and called "Greenbar" because it had shaded green bars across it – horizontal stripes alternating between white (no color) and green. The lines on the page aided the reader in tracking the printed data across the report, facilitating ease of reading and comprehension.

The entire business of Stat Tab was based around these processes, or jobs, consisting of coding surveys and other information, turning the codes into cards, and the cards into reports or check runs. Each job represented a project for a client. It was a thriving business in the pure art of data processing. To demonstrate how business environments have changed over the years, at Stat Tab, whenever we needed to run a job in the processing room, we had to schedule it through the Operations Manager, who insisted that we have a drink with him from a bottle of whiskey he kept in his bottom desk drawer. There were no restrictions about drinking (or smoking) on the job at Stat Tab during the whole time I worked there.

These first experiences taught me how data could be turned from qualitative information (e.g., open-ended survey questions) into processable data. I also learned the concepts of print layouts, the fundamentals of programming (in wiring the boards), how to sort data in diverse ways, and how to operate all the various IBM devices. I could key-punch and key-verify almost as fast as the best operators. In addition to that, I also acquired the skill of efficiently carrying an entire tray of IBM cards using just one arm while steadying it with my other hand. This involved managing well over 1,000 cards and smoothly transferring them from one tray to another or from a tray to the feed on a sorter or accounting machine. I became adept at performing this task without spilling the cards, well, most times.

At the end of the summer, just before I left, the company got its first IBM 1401-8K processor with a card reader/punch, printer, and tape drivers. It was a mid-range business computer designed for smaller organizations that required data processing capabilities. The IBM 1401 was the most advanced at the time Stat Tab got theirs. It quickly gained popularity and became one of the most successful and widely used computers of its time, with thousands of units being sold worldwide.

The person chosen to operate it was John Lyons, who would become a mentor to other programmers and me later. He introduced me to this first computer the week before I left. This gave me an excellent preview of where I was headed on my career path and solidified my career vision.

On my last day before leaving for my last semester, my boss informed me that the company wanted me to return after I finished my degree. I was guaranteed a job after college! That was more than comforting to a young man with a young family.

My Last Semester

When I returned to the campus for my final semester, one of the challenges I faced was getting enough credits to qualify for a degree, let alone a degree in math.

Getting a degree in Math from the College of Liberal Arts and Sciences required 120 total hours, two theory classes, and twenty hours in Math classes. I took two theory classes in the semester before my last semester, the first of my two Senior semesters. One, a calculus class in the physics department, which I passed. The other, Number Theory, I did not. This class was a "bear." It was taught from a textbook that was only a ½ inch thick, and the course only covered the first half of the book.

The teacher was also a very dull instructor, teaching over our heads to the three graduate students sitting in the front row. To shorten a long story, I failed the class, mainly because the instructor disliked two fellow students and me. We deserved his dislike for cutting out of his class whenever he was more than ten minutes late, which the school allowed us to do. It did not help that the class was held at 3:00 PM on Fridays on a campus known for its beer consumption. The professor failed all three of us, even though the other two were straight-A math majors. I went to his office to complain, but he was loud, intransigent, and argumentative in front of his office mate.

After a summer break, when I returned to school, I learned that the other two math students had stayed around campus, petitioned to be reexamined, and successfully passed the class over the summer. I was incensed. I did not even know that was an option, although I could not stay around the campus over the summer because I had to work and had to join my expectant wife in Chicago.

Soon, I headed into my final semester. If I passed all my classes, I would end up with 117 hours, three hours short. I had one theory class, which was one theory class short, and I only had passed seventeen hours in math, three hours short. If I could be reexamined in that three-hour class and pass, I would squeak by with all three minimums.

Unfortunately, I was out of money for college, and the University required me to finish only with courses taken on campus in Champaign, IL. If I did not finish this semester, returning for another to finish my degree was not an option. With no other options, completing my degree hinged entirely on the single class on Number Theory.

I asked to be re-examined. The professor who did not like me had left the university, but his office mate, who had witnessed the bad blood between the original professor and me, and now also disliked me, agreed to reexamine me. I took the exam and failed it.

I went home for thanksgiving and, on the return train ride, sat next to a lady who was an assistant dean. I told her my dilemma. She suggested that I go to the Head of the Mathematics Department and ask him if I could be reexamined again. (An ironic concept like Advanced Remedial Rhetoric). It was worth a try. Again, an angel helped me along my path.

I went to the Head of the Math Department and asked. He said he did not know if a second reexamination was allowed. He pulled a tome of university regulations and tried to determine if it was allowed. He said the book was "mute" on the subject. He handed me a permission slip and told me that if the Dean of my College approved it, he would have me reexamined a second time.

I went to the Dean's office, where there was a new Dean. I had gotten along famously with the prior Dean but had just argued with the new Dean the day before and did not think my prospects of getting the slip signed were good. When I arrived, an extensive line snaked away from the front of the counter where the Dean stood behind it, dealing like a mad man with student after student on their various issues. My hopes sank.

The secretary to the original Dean, who knew me well, saw me eyeing the situation and waved for me to come behind the counter. She asked me what I needed, and I told her. She said, "Watch this!" She took the slip from me, walked over to the harried Dean, banged him on the elbow, and stuck the slip in front of him. He signed without reading it. Again, I had been saved by another angel who helped me along my way.

I returned the slip to the Department Head. He called in the professor mentioned above and said, "I want you to reexamine this student in Number Theory again, and this time I want to see the results." Wow!" I thought, "That was heavy."

Hoping I was in the clear and thinking I only had to worry about learning Number Theory, I found a tutor and went to work preparing. Unbeknownst to me, the troublesome professor had more torment for me. He scheduled the reexamination right in the middle of Finals Week. He put me alone in the center of their largest lecture hall, sat me in the middle of the room, and then stood leaning against the doorframe glaring at me the whole time. I felt more pressure than ever before. I finished the test as best I could but did not feel confident. At the U. of I, inserting a self-addressed postcard into the test booklet was standard practice so the grade could be sent to you faster than the slow University processes. I then finished my finals for my other classes, which, feeling the burden of responsibility of having a family, I passed readily. Finally, I returned home, wondering if I would be granted a degree or live forever with an incomplete college career.

Five weeks later, a postcard arrived in the mail, bearing the words, "Passed, barely!" Even with that not-so-subtle final "shot," I remained unfazed. I knew my degree in Mathematics from the University of Illinois would be a key asset throughout my career.

What I Learned about I.T. in College

As you may know, what a student learns in college will often have little specific value after they graduate even if they go to work in their chosen field. And fewer than half of college graduates end up working in the field they studied. In the early days of computers, since, as I mentioned earlier, no computer science curriculums were offered in this field, only certain classes were relevant to where I was now headed.[2]

When I returned to college after my summer at Stat Tab, I took two classes that I thought would be relevant to my future work there. One class was "Advanced Statistics for Psychology," and the other was "The Theory of Computers." Oddly enough, both classes used the same textbook, *Mathematical Psychology*. The statistics class used a complicated electric calculating machine. This machine was like a souped-up adding machine that could take in a stream of numbers and then calculate averages, standard deviations, regressions, and even multiple regressions. The class taught me how to calculate different statistical measures step by step, and I found this knowledge useful when I returned to work at Stat Tab.

The other class on the Theory of Computers was difficult. The professor was almost unintelligible, talking in vague generalities. Half the class dropped the course after the first meeting, and another 25% dropped out of the class after the second meeting. Only eight remained. I needed the theory class for my degree, so I had no choice but to hang in. I decided my only hope was to take copious notes and, when the time came, spew them back to him at exam time. The class required a mid-term paper and the final.

I did exactly as I planned for the paper, regurgitating his lectures. I received an A+ on my paper. Sweet! I still have that paper, and it makes no more sense to me today than it did back then.

When the final came around, I still had not learned anything in the class but reviewed my notes a dozen times or more, hoping something would sink in. Upon completing the final exam, I felt I might even get my first A in college.

A week later, the teacher posted our grades on a bulletin board so his students would not have to wait for the University to send them out. I was surprised to see that I got a C in the class. I asked the professor how I could have dropped from an A to a C. He blithely said I had "tanked" the final, providing no details. To this day, I do not feel I "tanked" that test, but at that point, I did not pursue the issue because I had enough of a grade for my purposes.

In addition to taking these classes and carrying a full load of eighteen semester-hours, I also needed to work. For a while, I worked in a Mycology Lab (study of mold), sterilizing Petri dishes and doing other clean-ups. I also worked as a waiter in a fraternity for my meals. But the best and most relevant job I found was working on wiring diagrams for Illiac IV.

The Illiac IV was a supercomputer designed at the University of Illinois and built by Burroughs. The one-of-a-kind ILLIAC IV would take six years to complete at a cost of $40 million. The system was intended to have 256 64-bit floating point units (FPUs), each with its own memory and four central processing units (CPUs) able to process 1 billion operations per second.[22] Due to budget constraints, the project would build only a single "quadrant" with 64 FPUs and a single CPU. Illiac IV was a huge computer.

At the time I was there, the computer that the university was designing to be the fastest Supercomputer in the world, for its time, was being built under a government grant. The project manager hired me among others to verify wiring connections shown as lines on large blueprint diagrams. It was simple grunt work but paid well -- almost double per hour compared to my other college-town jobs.

On my first day, I knocked out the first batch in less than two hours and returned them to the Supervisor. When I handed my work to him, he made an odd face that I could not interpret. The next day, when I finished again in less than two hours and turned in my work, he told me to close his office door and sit down. He then lectured me on being too efficient and doing the work too quickly. As he explained, it was a government grant meant to be "milked" for all the hours originally planned, so that the government would not take back unspent money at year-end.

My farmland work ethic, Catholic upbringing, and family principles would not allow me to sandbag on the contract, so I quit. At that point, I had enough money to make it through.

Back to Stat Tab-Early Growth

Following my graduation in January, I went back to Stat Tab, and true to their promise, they rehired me. The pay was $600 monthly, $7,200 per year, plus medical, which was GREAT! I felt like a grown-up. It was more than enough to live on, so we found a new apartment on a shady tree-lined street in Rogers Park on the north side of Chicago. When we signed up for the apartment, the property owner said he would do us a favor and not require a lease. We took his kind offer at face value.

Little did we know that he only wanted a tenant for three months. When he told us we had to move out, we were angry, but that same day, another apartment across the street became available. It was bigger, newly remodeled, and at a price we could just afford. We lived there for four years, and the landlords, a "bear" of a man and his sweet wife, never raised our rent.

Before I arrived at Stat Tab for my first day back, I was informed that the company had moved from the Monadnock building to offices in the Northwestern Train Station. The offices were not near enough to the trains to hear their noise and were like any other work location. The offices included a large computer room to house the IBM 1401 with 8k and its new neighbor, another IBM 1401 with 16k memory, and their supporting equipment. It also had a raised area for the

Keypunch/Keyverify operators. At the end of that area were a dozen individual offices. I was given my own small office, with just enough room for a desk and one chair, which made me feel important and gave me quiet to concentrate when coding. This is what an IBM 1401 looked like.

My boss was a gentleman named Wallace Rounds. He smoked a pipe and had a Ph.D. in Mathematics but did not use the title, Doctor. He mainly worked as a Systems Analyst, designing the sets of programs, their relationships, and the processing needed to complete a customer's job. He doled out the programming work to four programmers.

He said he would get me business cards and asked what title he should put on my card. I suggested, "Mathematician." That seemed appropriate, given my new degree in Math of which I was so proud. He argued with me a little since he was a Ph.D. Mathematician but acquiesced. When I got the cards a week later, they read: "Mathematician-Programmer Trainee." He had diluted my pride down to a dose of reality. In the end, I did not mind because it was accurate, and I was not one to stand on formalities.

I had learned some Fortran II in college. So, my first assignment was to write a program in Fortran. I also learned Autocoder, an early programming language, on the job.[4] This is when I truly began my education as a programmer. As mentioned above, our computers were two IBM 1401s – one with 8K memory and the other with 16K.

It was easy to fill up a 1401's storage with 8K of programming code and data, so we had to learn special techniques like "overlaying." To overlay, we would write programs in segments with the last instruction of the first segment executing an instruction to read in the next program segment. This would cause the second segment to overwrite (overlay the same space in memory) as the first segment. If the program had to return to the beginning, which sometimes it did, an instruction to read in the first segment would be executed. The segments were stored on a tape drive, which would spin back and forth as the alternating segments were swapped in and out.

We also learned to "double buffer," a technique to speed up data input from peripheral devices like tape drives or card-readers. Reading data from a peripheral device was slow compared to processing data in memory. The standard programming method for reading data consisted of two instructions: a Read instruction and an instruction to verify that the Read had been completed successfully. Executing these two instructions repetitively in a looping block of code caused the program to run slowly. The Double Buffering technique involved starting the read process, then starting another read process, then checking if the first one was finished, then starting another read, checking that the second read was completed, and so on.

You can envision this method by picturing yourself frying eggs: crack one into the pan. Once the first one is nearly finished, crack another one into the pan. When the first one is almost done, remove it and start another. Repeat. This overlapping of the reading-in of data took advantage of the lag time it took for data to be delivered from the tape drive and speeded up the programs noticeably.

In addition, we would write blocks of code into subroutines, now called functions, when they are built into the coding language, so that we could use them repeatedly within the same program or in other programs we might code later. Many of the functions we wrote back then are built into programming languages today, like sort routines, math functions, rounding decimals, etc.

So, for example, we learned that there are different ways to sort data, with some being faster than others. Sorting a single column of data is more straightforward than sorting a data matrix, where one column is the "major" sort field, and another is a "minor" sort field. Think of creating order within an order. For example, a major sort field might be the residence city, and a minor sort field might be the street name. Each sort field might be sorted alphabetically, in ascending or descending order. All these details we packaged into subroutines or functions. Programmers today still build off the vast amount of work done by the programmers who came before them.

The migration of a sizable portion of programming code into canned functions, now widely accessible across programming languages, is a prime illustration of the constant march of "change" and how I.T. steadily evolves.

The learning environment we enjoyed was priceless. In my early days of programming, changes were happening so rapidly that IBM issued weekly reprints of instruction manuals for each programming language and each computer and peripheral device (the card punches, printers, and tape drives). I read every manual they issued on all the various programming languages, and hardware we used, and the accumulation of manuals in my office was voluminous. Printed on very thin paper and printed in small font, the manuals sometimes would stack up from my office floor to higher than the top of my desk. Sometimes I had two stacks to read. Nevertheless, these manuals were the only way to learn to program and keep up with the constant language and hardware changes. I taught myself to speed read them, recalling the last issue, and scanning for what was different.

Stat Tab had only four programmers, and we stayed busy. Tutoring by my fellow programmers was another way I learned programming. John Lyons, the nice guy who introduced me to their first 1401 at the end of my first stint at Stat Tab, was a specialist in multiple coding languages, including machine language (codes the computer understood directly) and IBM's Assembler. Fortran code, my expertise, had to be compiled, meaning run through a particular program that turned my Fortran code into Machine Language.

Another programmer, Dick Nastav, specialized only in COBOL, which also required compiling. COBOL used lines of code that read like English but had a lot of quirks. The fourth programmer, Dave (last name forgotten), specialized in RPG (Versions I, II, III, and later IV), a business language that worked off parameter cards rather than compiling. The one thing I remember about Dave was he had joined the U.S. Navy ("to see the world") but learned in the middle of basic training, to his surprise, that he was transferred into the Marines. Prior to being informed, he did not even know the Marines were part of the Navy.

We all learned to program in each other's languages and the basic machine language (Autocoder, Assembler, and the core machine coding). I also learned Basic, a lot like Fortran but easier for small jobs. We not only cross-trained in each other's coding languages, we received overflow jobs from each other when workload demanded.

The four of us would work all week feverishly, and then, around 3 PM on Friday, John would show up with a couple of six-packs of beer, which would be the end of programming for the week. Sometimes we would have to work nights or weekends if there were long-running jobs or tight deadlines.

The work environment was terrific for learning. But, as programmers, we were prima donnas with the run of the computer room and in directing the work done by the machinery, computer, and keypunch operators. I remember a computer operator whose forehead was all red and peeling. When I asked him about that, he said it was the "damned computers." When I probed him further, he said it was because the computers never made a mistake; every mistake was his fault. He was all stressed out from this new type of work – flawed humans running flawless computers.

We programmers often "played" with the computers, trying different things, arguing with each other about what was happening inside the computers, and digging into the guts of how computers worked. One discussion we had was about how a three-card deck provided by IBM worked. When the three cards were put in front of a stack of cards in a particular order, and the stack ran into the card reader, the computer would print out the cards' contents. If we reversed the second and third cards of the IBM mini-deck and ran them in front of a stack of cards again fed into the card reader, the computer would punch out duplicates of all the cards in the stack. This was voodoo to us. So, we decoded the punches in the three cards to learn how the program they represented worked. Once we did that, I wanted to know more. I wanted to see if I could reduce the program to just two cards (only reverse the two cards to do the same thing). After working hard on this self-invented challenge, I succeeded in reducing the code to two cards and one column on the third card. Unfortunately, all I learned was that squeezing out that last instruction was not possible.

We also endured frustrating incidents such as accidentally dropping an entire tray of meticulously sorted cards while balancing them on one arm or encountering tape drives that inexplicably started rotating in reverse.

When a tape was running (being read) on a tape drive, it threaded through two vacuum columns, one on each side of the reel, to keep it taut and running smoothly. The tape drive looks like a long, narrow letter W, with the reel at the top and center. However, occasionally, the reel would get a mind of its own and start unreeling backwards. The accumulation of the tape's weight inside the vacuum tube would hasten the unraveling of the tape. If you did not catch it immediately, the entire tape would end up in a pile in the left vacuum channel. The piled-up tape would get creased in the process, often making the data on it unreadable. The only option then was to start the entire project over from the beginning. This experience was like having your computer crash just before you did a "save" after doing several hours of work. Saving your work was an essential skill we needed to follow consistently. Now programs often do saves for you. I still do not trust them to save my work.

One thing I learned while "playing" between assignments at Stat Tab was the answer to this question which very few if any, programmers even today would know: "What is the first instruction executed inside a computer when it first boots up?"

The answer is: "Read the next instruction." That process is called "bootstrapping," The only way a computer can get started when turned on is to do this first instruction. Try this question on your favorite IT person and see if they know the answer. I would wager big bucks that they do not.

In 1966 my wife and I had a second child, a daughter, Allesa Mary.

Naked Honeywell 200

Before long, the company acquired a computer from Honeywell. This new computer was physically large and had six tape drives, 512K memory, and standard peripherals -- a card reader/punch and a printer. However, it lacked the most important thing -- an operating system. Without an operating system, the Honeywell 200 was good for no more than a giant door stop.

However, my boss had a job that needed the power of the Honeywell 200 (H200), which he delegated to me. So, before I could program this assignment, I would need to write an Operating System, much like Bill Gates and Paul Allen did in Bill's garage. And, of course, I had no idea where to start. So,

I asked my boss if he could get me help from Honeywell on how to program their computer. He said he would. Meanwhile, I was studying their manuals. This is what the H200 looked like:

Soon after my request, late one Friday afternoon, a man walked into my small office and sat down. He introduced himself as the guy from Honeywell. He was Germanic with an accent. He bluntly asked, "Vhat is your problem?" That set me back, and I had to think hard. He sat patiently while I thought. Finally, I blurted out, "I do not know how to program." There was a raw truth. I had been writing programs but not operating systems and my coding, like my writing, was not as well organized or as efficient as it should be. I expected him to stand up and walk out of my office, but instead, he said: "Then I vill teach you," and he did just that.

He came to my office around five PM for at least an hour every couple of days for weeks and guided me in learning programming in general. I then took what I learned from him and built an H200 Operating System (OS). In about six weeks, I had completed the OS long before the company provided an official one of its own. This meant writing Driver software for each attached peripheral, plus the logic to read in programs and run them, and more.

I then started writing my assigned program, which would utilize all six tape drives to process and sort down a large amount of data from a drug company to prove their drug was effective (it was not).

In 1967, I took the GMAT and applied to the University of Chicago's night school MBA program at their old location on East Delaware Place. I did it mainly because Stat Tab would pick up as much as 75% of the tuition based on my grades. (A= 75%, B=50%, C=25%). This motivated me to do well. I was working 50–60 hours per week, married with two kids, and stretching to make ends meet. I still had student loans from undergraduate and potentially new ones from graduate school. I figured the investment, energy, and money would be worth it and I could get a better-paying job with an MBA from a prestigious business school.

Despite performing well on the GMAT due to being "test-wise," my admission was uncertain because my undergraduate GPA was just above a C, as previously mentioned. Additionally, I was required to submit an essay, making my admission even more uncertain. My dyslexic brain never did well in writing essays because writing needs to be organized.

Nevertheless, I did my best and got advice from others before turning it in. My GMAT score got me in the door for the mandatory personal interview, which seemed to go okay. Despite my best efforts, my essay left my admission status uncertain. As a result, they invited me for a second interview. My acceptance hinged on that second interview.

In preparing for it, I reflected on my experience with the Assistant Dean in charge of ROTC at U. of Illinois. I thought ahead as to what questions he would ask me. Finally, I realized the critical question would be, "What was my motivation for getting an MBA?" I knew saying I was in it for the money would be a mistake.

During the interview, halfway through his questioning, he popped the question I anticipated: "And why do you want to get an MBA from the University of Chicago?" I answered that I wanted an MBA to learn all I could about business. That was the answer they were looking for, and I was accepted. Again, I had squeaked through.

We lived in Rogers Park and did not own a car, so I got around via public transportation such as buses and the "L." Unfortunately, my closest "L" stop to the MBA school was several blocks away, and on late blustery winter nights after class, the walk was uncomfortable at best. A fellow student I had gotten to know through being in three classes together was Willie Davis, MBA'68 (Died, Spring/20). Willie was a defensive end for the Green Bay Packers. He was a ferocious football player who would be voted into the Hall of Fame, among other accolades. I remember the contrast between the gouges out of his knuckles every Monday and his mild nature. I never met him on the field, but he was one of the nicest men off the field I had ever met: calm, collected, and even gracious. He was about ten years older than me and seemed to have brotherly kindness toward me.

On those cold winter nights, when I was tired and anxious to get home, he would drive me to my "L" stop. He told me he and his wife had gotten big, beautiful Lincoln Continentals for his doing a commercial. I thought that was so cool. When we graduated, he bought a Budweiser franchise in California (great timing, no doubt) while I went on to build large global computer systems. I never got to thank him after we parted ways, but I will always consider it an honor to have known him briefly through our discussions during those short car trips to my "L" stop. Meanwhile, my wife would keep our kids up until I could get home so I could read them a story before they went to sleep. This was a time of intense learning and hard work, only possible with the support of an equally committed and hard-working spouse who took in typing while she raised our two children. I must also compliment the University of Chicago for creating and operating an educational program that for the first time in my life captivated and motivated me to excel.

My First IBM 360

Soon after Stat Tab received the Honeywell 200, they added one of IBM's first model 360s to their stable. They still had the two IBM 1401s. So, we were writing programs on all four computers. IBM's CICS came out about the same time. CICS stands for Customer Information Control System and is a tool created by IBM to help computers manage high-volume transactions and applications. It was another language we would learn. It is still used today.

The IBM 360 was notable because it transitioned smoothly from the 1401s and came with the usual peripherals except for new storage devices -- random-access disk drives. We thought they were exciting. This was our first exposure to random access, and initially, I was the only one of us programmers who took on that challenge. Here is what its console looked like:

These drives came with removable "sets of platters." These "sets of platters" enabled the read/write arms to move like fingers on a hand, in and out. The read heads functioned akin to the arm of a record player that gets bumped, causing the needle to skip across the record and scratch it. In this case, the platters were not scratched; only the embedded bits were flipped from zeros to ones or vice versa. The device moved in two directions simultaneously – rotating the plates and moving the read heads in and out. They looked like:

This style of storage device has a unique function. It enables programs to access data much faster, resulting in faster program execution, particularly with large data files. In contrast, accessing the final piece of data on a tape required the entire tape to be scanned. With a rotating random-access disk, the head could simply jump to the end of the file. However, the kind of programming I did also used a few different instructions, which were not supported on IBM's 360's operating system when it first came out.

Many programs I wrote utilized Column-Binary, which I described earlier, where each column of the IBM Cards could have multiple punches. When I tried to run my programs on the IBM 360, I discovered it was not supporting column-binary on the card reader or the card punch, although the reader/punch peripheral had the capability.

I requested a copy of IBM's Operating System, which was referred to as DOS (Disk Operating System) at the time but was later renamed to OS/360. While it was not well-known then, DOS gained popularity later when IBM began selling it as the operating system for their desktop computers. To my surprise, IBM did provide me with a copy of the source code. I modified the DOS code to accommodate my column-binary needs. My prior experience writing the operating system for the Honeywell 200 gave me the know-how to make the changes needed.

One of my proficiencies by the nature of my primary programming language, Fortran, was programming jobs that incorporated Statistical computations in their designs. Over the course of a year or so, I wrote computer programs to perform every statistical calculation commonly used in business and science. Then, with the arrival of the IBM 360 and its random-access drives, I built my own "subsystem" that would apply any statistical computation to a stack of IBM cards that followed a few Parameter cards placed in front.

I stored the routines that did the calculations on the Random-Access drives and then only had to create a few parameter cards to put in front of the cards with the data to be processed. And, for my amusement and so anyone could use my subsystem, I wrote the programs to interpret English language instructions on the parameter cards. For example, I could punch into a parameter card to "Add columns 4-6 to columns 7-9 and compute the average." I was intrigued by the concept of simplifying programming by using English words instead of code. By enabling non-programmers to operate this subsystem, the company would save money and significantly speed up the process of running statistical jobs.

While building this system, I heard about a product being developed elsewhere called Statistical Analysis System (SAS). I asked for and received a copy of SAS and tried it. It could have been better, and I found a few inaccuracies in the early product, so I continued developing my own. SAS would later become better -- more comprehensive, entirely accurate, and highly successful.

My First Sales Position

You would think that my hard work, technical expertise, and ingenuity would have resulted in lots of praise, recognition, and rewards. Well, it did not. About this time, I was offered and accepted a "promotion" into Sales. I was known for being clear when explaining complex data processing subjects to non-technical people, so I presume management felt this would translate to selling.

I did have sales experience of a sort. In high school, I had a part-time job selling women's shoes in a Burt's Shoe Store in Mount Prospect, IL, one of three Edison Brother's Shoe Company three brands. Bakers and Chandlers were their other brands. The Edison Brothers had a standard sales approach. For example, you always showed a woman five pairs of shoes – the one they asked to see, another color in the identical style, another similar type of shoe in the requested color, and two shoes they might not have considered. Their system worked very well. We often sold more than one pair. We also sold matching purses and accessories like add-on bows and shoe polish. We worked on commission (minimum wage as the floor), with higher commissions for handbags and even higher for accessories.

My shoe sales job allowed me to find a job when I moved to downtown Chicago to go to college. I walked into the Chandlers on Michigan Avenue and was hired on the spot because I was already trained. I was given the enviable additional responsibility of dyeing white silk or white fabric shoes for ladies who wanted shoes that matched the color of their dress. I say enviable because those customers included every new Bunny from the Playboy Club down the street. I received this honor because I was the only male in the store, other than the manager, who was not color-blind.

The manager told me I could work whatever hours I wanted. This meant I could choose to work when the store was the busiest, helping the manager cover his peak times, and it was when I stood to make the most money. When the minimum wage was $1.00 per hour, I averaged $7.50 per hour, working only 20 hours per week. It was a perfect job for a college student paying his way through school and renting a one-room apartment. And this taught me a bit about retail sales but little about corporate sales. So, I was unprepared for my new Stat Tab job.

I was assigned to report to Jerry M., the only salesperson on staff. When I was added as a second salesperson, he feared that his commissions could be diluted. His first task was assigning me a Sales Territory, so he chose the entire U.S. south of E. Lake Street in Chicago. The only problem was that all existing customers' offices were north of Lake Street. So, my territory would be 100% cold calling. Besides the Loop area, I pursued leads as far away as Evansville, at the southernmost tip of Indiana.

Unexpected Major Talk

One funny thing that happened to me in this job, the Realty Advisors of America (REA) asked our company to have someone give a 30-minute overview about computers at a meeting of their Regional Directors. Computers were a new topic for many businesspeople. Our company president asked the sales guy, Jerry M., to give the talk. Jerry told our local manager that he was unavailable on the day of the REA meeting and asked if my old boss, Wallace, could give the talk. Wallace asked if I would do it. I had never given a formal speech before, so I agreed to do it – for the experience. I had acted in a couple of plays and a musical in high school, so I thought I would be okay regarding stage fright. I worked hard to prepare my 30-minute talk, including preparing overhead slides.

One day before the scheduled talk, Jerry came and told me REA was now asking for a one-hour talk. So, I spent the rest of the day expanding my talk and adding slides.

The next day, when it was time to leave for the talk, Jerry showed up and said his conflict was canceled at the last minute and that he would be accompanying me to the meeting. I never believed he had a conflict, but rather feared public speaking, and was planning to take credit for any sales. But unfortunately, it was too late for me to back out.

On arrival, we were directed to a door with a small round window. It was the doorway to a nightclub. Looking through the window, my heart sank as I saw a dark room with over 100 people inside and a stage with spotlights, two microphones, one for recording the talk, and a lectern. This was a far more formal event than I was expecting. A young assistant came out and, in the most amicable terms, asked me if I would kindly extend my talk to cover an hour before lunch and an hour after lunch. I was sure my 30-minute talk, which I had stretched to 60 minutes, would not stretch to another hour. I could have killed Jerry on the spot!

Stage fright hit me hard as I was introduced and stepped onto the stage. I stumbled through the first hour, drenched in sweat and anticipating the metaphorical hook to yank me away at any second. I felt like the most inept public speaker alive.

When we broke for lunch, I joined a table of attendees. They were cordial, seemed fine with my talk so far, and showed lots of interest in learning more about the subject. My confidence skyrocketed, and my second hour went swimmingly because I realized how little they knew about computers and how much I did know. However, I still wanted to kill Jerry.

Coming to an End

The rest of my time in sales was unsuccessful and unfulfilling, so I will skip the rest of those gory details. Except, I did have one client who loved to go to lunch and watch the Chicago Cubs play on TV. I just had to hang out with him for the entire game. And sometimes, he would join me when the company's IBM sales representative would give me box seat tickets for the Cubs, right at third base – the best seats in Wrigley.

Meanwhile, I was devoting some time to finishing my Statistical Subsystem. I estimated it was only two weeks from being completed when my Sales Boss told me to stop working on it. No matter how much I pleaded and tried to get my Sales Boss to understand, he could not grasp the system's value or potential and stuck to his order for me to abandon the project. A project that could have made Stat Tab millions of dollars was thrown into the trash bin that day. In hindsight, I should have kept what I had done and started my own business with it, but I lacked the capital, vision, and the confidence to strike out on my own with a small family to house and feed, and school to finish.

I soon graduated with my MBA and started looking for a new job. The owner of Stat Tab, I would hear later, was very upset with my Sales Boss and his boss, for not introducing me to him, as he had job opportunities at his headquarters.

The industry career path in those days was from programmer to senior programmer to programming Manager or from senior programmer to systems analyst to senior system analyst, then to programming manager, and so on. I moved from 'programmer-trainee" to senior programmer at

Stat Tab in three years. I began looking for a systems analyst position or a manager of programming position to utilize my recent MBA education.

Opportunity Missed?

While I quietly searched for a new job, our group of four highly skilled programmers at Stat Tab met with a potential investor off campus. We discussed the possibility of starting our own data processing company, like Electronic Data Systems (EDS) which Ross Perot acquired from General Motors and led in Plano, Texas from 1962 to 1988.[23] It was a formative time for the four of us, but there were hitches.

First, none of us were salespeople, and we would have to hire that skill, which was possible. Second, the other three programmers were single, and I was married with two kids. So, the risk tolerances were quite different between me and the others. Lastly, the investor wanted to start as the majority shareholder, which did not seem fair to the four guys bringing all the skills to the venture. We also did not know the investor well, so ultimately, we decided against it. Also, in hindsight, our most senior programmer and our potential leader had no experience in setting up or running a business.

My job search went quickly. I first considered taking a job at Playboy Magazine after I had helped build their mailing list at Stat Tab, but when I went to interview, I saw that the workstations were tiny little spaces along a wall. I already had an office and had no desire to take a step backward. This exemplified the reduced significance of programming within their organization.

My next opportunity boiled down to two competing options. One was the start-up of a new credit card company, and the other was United Airlines. The Credit Card company was going to be called VISA. I liked the guy who would be hiring me at VISA, but I did not hear from him for a couple of weeks after my interview, so I concentrated on United Airlines.

With United, I would be a Senior Systems Analyst working on a new Reservations System that would run on a three-processor Univac Unisys System called Unimatic. United Airlines was the more attractive offer because my family could fly free on standby or at a significantly reduced fare for reserved seating.

As my wife and I had only flown once in our lifetimes, and we liked the prospect of traveling, this made our decision. I would not have my own office for another three years, but I was okay with that trade-off. And there were new computers to explore – multi-processors.

Three weeks after I started at United, the Visa guy called, sheepishly asking if I had taken a job yet and apologizing for his delay in getting back to me. He still wanted to hire me. I would have been Visa's first I.T. guy. However, I had committed and was not inclined to change my mind, so I declined. I will never know what kind of a deal I might have struck with him, maybe substantial stock options?

CHAPTER 2
UNITED AIRLINES

Unimatic

Unimatic was a major new development project for United Airlines. Programming work had already begun before I arrived. The system would include all the primary United Airlines functions, Flight Operations, Crew Scheduling, Reservations & Ticketing, and others. It was a grand vision – one system to run the whole airline.

The computer was built from three inter-connected Univac Processors. In addition to the three massive processors, the system was surrounded by giant storage drums – think of an oversized barrel on its side in a rectangular case with a window in the front top. These were random access devices, spinning at high speed with arms that ran back and forth to find data in three dimensions. Instead of card readers reading punched cards, the input devices were Cathode Ray Tube (CRT) terminals that people typed on. We wrote our code on coding sheets, and another department typed the code into the computer. We revised on Greenbar printouts.

The system was to be written in Fortran V, my most familiar coding language. However, the Univac processors operated radically differently. All data in Unimatic, including coding instructions, were stored as "words," meaning a block of octal number pairs, such as 12, representing the number 10, in our more familiar number base of 10. You may or may not have learned about different number bases or may have forgotten, so here is a short primer. Feel free to skip over it if you are not technically inclined.

You can convert an octal number into our familiar decimal numbers by breaking the number into single digits, counting the two digits in the octal number 12 as one eight plus two ones or 1 x 81 plus 2 x 80 or the number 10 in decimal numbers This works the same with binary. For example, the binary bits 1010 mean one 8 (as in 23) plus no 4's (22) plus one 2 (21) plus no ones (20), or 10 in decimal. Each position counting from the right in both cases above is the exponent of the base number. Clear as mud, right?

Octal numbers run from 00 to 77, skipping numbers with either 8 or 9 and these translate into decimals 0 to 63, or 64 possibilities. Octal 77 means 7 times 8^1 (=56) plus 7 times 8^0 (7).

These 64 possibilities were used to represent the alphabet (26 lowercase letters and 26 uppercase letters), the numbers (0-9, special characters, and a few "control characters," like carriage-return-linefeed, space, etc. So, for Univac, in our example, 8,000 words of storage could hold 512,000 characters. Unfortunately, packing code into Univac's octal words made programming difficult because we had to ensure we did not leave empty bytes in words. Otherwise, it would consume too much memory and slow down processing.

When I arrived, they sat me in a large cubicle of eight desks. We all faced west; mine was in the front, so I looked at a blank carpeted wall in front of my desk. I soon learned that the programmer behind me was one of the most knowledgeable and one from whom I could learn quickly. Initially, I turned my chair around to bug him so often with questions that he became perturbed with me. It was not long, however, before I found my footing and began to "fly" on my own.

I soon became a Lead Programmer, supervising a group of five other programmers. Informally, I also became the unofficial Master of Testing. When other programmers in other sections felt their code was solid, they would invite me to test their code, and I inevitably found bugs – exasperating them time after time. My skill came from my intimate knowledge of how computers worked and from my years of programming. I would look for the "extreme corners," such as entering letters in number fields, entering negative numbers in number fields meant only for positive numbers, entering special characters where they did not belong, entering inputs longer than the data field was expecting, and more. Unfortunately, many programmers tend to check for only the obvious.

The One That Got Away

My role included hiring at times. I always tried to hire without bias. There were few women or minority programmers in the general marketplace, even fewer in specialized industries likes airlines. Once at United Airlines, I interviewed an African American gentleman who interviewed very strongly. He was articulate, had all the correct answers to my interview questions, and passed our programming test. I was prepared to make an offer but suddenly realized I had not seen any African Americans in the entire Executive Offices (EXO) building.

Concerned about what I might learn from my boss, I approached him anyway to see if unwritten bias was happening around me. He said there was no problem as far as he knew and to do whatever I felt was right. So, I made a solid offer to the young man. Strangely, I never heard from him again. I later wondered if the sharp young man might have been an attorney testing the company to see if we would make an offer to a highly qualified minority candidate. I know that sounds cynical, but I could think of no other reason why a generous pay offer, compared to the market, with free flights, medical, and ample paid vacation time, would not have been accepted or at least acknowledged. He was polite in the interview, so I expected at least a polite decline, but I heard nothing from him either way. One never knows in the hiring business.

Special Consultation

We were occasionally called into meetings with Unimatic managers to discuss technical design issues. I remember one where the discussion concerned whether the system should operate its communications as Full Duplex or Half Duplex. With Full Duplex, every message sent across the data lines from the computer to a computer terminal, for example, at Pittsburg airport, would be acknowledged with a response confirming that all the bytes (characters) sent out were received. In other words, if 120 characters were sent out, a message was sent back with the news that exactly 120 characters were received. For this to happen, a field had to be appended to the end of every transmission with the count of the characters initially included in the message. The receiving program would count the letters and check their count against the last field. If the receiving program had to request a repeat transmission, retransmission would add processing load on the computers when the count did not match this checking. This checking was deemed necessary because sometimes characters got dropped in transmission.

Of course, United's system would be transmitting millions of transactions daily, and processing capacity was already limited despite the power of the Unisys computers. On the flip side, Half Duplex was simple and easier to program. With Half Duplex, the system would send the message out, hope for no dropped bits, and live with the error rate, assuming it was low enough not to be a problem. No one knew, however, what low enough should be. With me being one of the voters in favor of Full Duplex, it carried the day. I figured that the telecommunications lines would get faster over time, so the extra traffic load would not be an issue long-term, which is what happened.

Ten Pounds in a Five Pound Bag

After six months, my team was warned that the Reservations System might be offloaded from Unimatic because the combinations of Applications were becoming too much for our Unisys computers. So, to stave off the removal of our hard work, I developed a system where our reservations programs would operate with "parameters" fed into a core "interpreter."

Here is how an Interpreter works, for those who are tech inclined:

An interpreter is a method used to convert high-level programming languages into machine language that a computer can execute. Unlike compilation, which transforms the entire program into machine code before execution, interpreting reads and converts the high-level programming language into machine language one line at a time as it is being executed.

When using an interpreter, there is no creation of a separate machine language program. Instead, a master machine language program sits in memory, ready to interpret the code. The interpreter reads each line of the high-level code, converts it into machine language, and then sends it to the processor for immediate execution.

The advantage is that interpreting does not require the upfront compilation step, allowing for quicker development cycles as changes in the code can be immediately executed without the need for recompiling the entire program. Additionally, interpreting may use less memory because there is no need to store a separate machine language program. However, interpreting can be slower in terms of execution speed compared to compiled programs. This is because the code is converted into machine language on-the-fly during execution, which introduces additional overhead.

In summary, an interpreter converts high-level code to machine language line by line during execution without creating a separate machine code file. It enables faster development and potential memory savings but runs slower than compiled programs.

So basically, an interpreter translates your code line by line—like a patient but slow waiter who insists on explaining every ingredient before serving your meal.

As I understood the issue with the Unisys processors, there were too many programs consuming their memory and overloading their processors. The processors might be sufficiently fast if the amount of code were reduced. So, I designed and programmed a core program of our own, and then all the reservation programs were built as a sequence of parameters stuffed into a half dozen octal "words." This approach was like the statistical system I had developed at Stat Tab. It only took my team three weeks to rewrite our Reservations System to utilize this approach. When tested, it ran like "lightning." It was incredibly efficient.

However, our work was not enough to free up all the space required by the other functions in Unimatic. There was another alternative, so management decided to offload the Reservations System. The company decided that the new system would operate at a new computer center to be built in Denver, CO, on IBM computers. Up to this point, I was known as a top-level programmer and, like at Stat Tab, someone who could explain complex issues to users in understandable terms. This skill got me noticed by the head user named Barbara Allen.

When this decision was made, my manager, Leo McGrane, presented me with three options. I could remain with Unimatic and work on the remaining applications, Flight Operations, or Crew Scheduling with a promotion to manager. Or I could transfer to Denver and be promoted to programming manager on the new system using a programming language called ACP, Airline Control Programming, which I did not yet know. Or I could join a new Marketing Division group, on a lateral move, to help define precisely how the new Reservations system would work.

The plan was to purchase a copy of the Eastern Airlines Reservation System, which they had modified from a copy of American Airlines' Sabre system. Barbara Allen would head up the group. I had already been working with Barbara and respected her. Each of my three options had pros and cons.

Leo was a respected, easy-going manager, so I invited him to dinner to help me think through my options. By the end of the meal, my decision was clear: make a lateral move and set aside promotions for now. Why? Upward mobility. The "user side" of the business offered more paths to higher-level positions than the technical side. As a technical expert among non-technical users, I could have a more significant impact.

Meanwhile, technical organizations have a broad base with few positions at the top, limiting advancement. Plus, programming success doesn't necessarily prepare one for management. Moving to the user side gave me flexibility— and that decision proved correct.

Apollo

Going to work for Barbara Allen was one of the best decisions in my career. United Airlines was a very "male-oriented" organization prior to the 1960's. Pilots were all male; ground crews were all male, and reflecting the company's military bearing, all the walls were painted with the Air Force color, three and a half grey. Barbara was once described as "She thinks so much like a guy that you forget to hold open the door for her." That was a compliment in 1969 but might be considered an insult today. A better way to describe her would be to say she could outthink most men around her and charm them into doing things her way with her wit and intelligence.

Her team was simple; she had three staff members from the technical side of the business and three from the Reservations Operations side. While we all had the same title, Applications Planner, I filled the informal role of Technical Lead because I had technical experience on the prior Reservations System. If I recall correctly, Joe Pantano was the informal lead on the team's Reservations Operations side.

How to Develop Large Systems

Our Technical Application Planners were essentially systems analysts who could communicate clearly with users and understand their needs. Our User Applications Planners used their experience as users to ensure that the system would serve the user's needs.

Here is the distinction between Systems Analysts and Programmers if this interests you:

> A systems analyst and a programmer have two distinct roles in the field of computer science and software development, although there may be some overlap between the two positions.
> A systems analyst is responsible for analyzing business processes, systems, and requirements to identify areas where technology can be used to improve efficiency and effectiveness. They collaborate closely with stakeholders, such as

business managers and users, to understand their needs and requirements and then translate those requirements into technical specifications for software development. Systems analysts are also responsible for designing and testing new systems or applications and evaluating existing systems to identify areas for improvement.

On the other hand, a programmer, also known as a software developer, is responsible for writing and testing code to create software applications. They take the technical specifications provided by systems analysts and turn them into working software programs. Programmers also maintain and update existing software, fix bugs, and ensure that applications run smoothly.

In summary, a systems analyst focuses more on the business requirements and overall system design, while a programmer focuses more on coding and implementing the software. Both roles are essential for the development and maintenance of software systems.

Functional Specifications became the key to successfully building large and complex computer systems. Our team under Barbara Allen developed this concept, and I further refined it later, during my years at Resort Condominiums International (RCI) and in subsequent jobs. The idea is to compile all the necessary information in a single document that a programmer will need to write each program. This document specifies various elements, including inputs, data processing, data elements, their relationships, storage retention requirements, key sort fields, calculations, sorting, outputs, security needs, and other pertinent details.

The Functional Specifications document serves as the foundation for the User Procedures document. The User Procedures document outlines how users will utilize the program and its impact on their job and includes sections for training, testing, and administration.

Together, these two documents facilitate a streamlined review, understanding, and approval process by managers to ensure their alignment of technical requirements with business requirements. It is a far superior process compared to reviewing programs developed without such documentation. Furthermore, this approval process ensures that the responsibility for the system design is shared with the users, allowing for tracing later issues back to the foundational documents. Additionally, these documents guiding the resulting development help to enforce accountability with clearly defined responsibilities at every step of the process.

This process offers multiple advantages, including pre-coding review of programming requirements and identification of flaws, ensuring programmers have a clear understanding of what to code. Additionally, the collection of documents defines the project scope, facilitating prompt issue resolution and easy reference throughout future project stages.

Later, when issues arise, and they always do, the question can be easily answered when the team is asked, "Was this need included in the original scope or not?" As I will expand on later, there are three dimensions to completing a large computer project: 1. Scope, 2. Budget, and 3. Deadline. Management can fix any two of these three variables but not all three. It is important for senior management to understand this concept because it protects the programming team from unfounded criticism and "scope creep." Although widely recognized today, this set of three constraints was initially a revelation to senior management when I presented it as a countermeasure against scope expansion after programming work had already begun.

In any project of size, arguments arise surrounding the original scope, and revisiting the pertinent Functional Specification usually settles any arguments. During the project's development phase, change management must be tightly controlled, as I will explain in more detail later.

After the Functional Specifications and User Procedures are signed off, they go to systems analysts and programmers to code, and the Procedures, once signed, go to Training Developers who create the user training classes. Training Developers and trainers are two different skill sets that should not be combined on a large project.

By following this disciplined approach, I successfully led four global, real-time system development projects – Apollo (now Galileo), FAMIS, RCI, and DVC - to successful completion within scope, on time, and within budget.

I designed a comprehensive System Development Process diagram laying out the steps involved in the design and development of any large-scale computer system, including creating a manual that describes the requirements of every step and providing forms and models for each step. I used this manual to train new development teams each time I moved to a new project.

Origin of Apollo, Later Galileo

While gearing up, we traveled to Eastern Airlines to be trained on their Reservations system. Surprisingly, Eastern Airlines had the legal rights to resell the system to United even though its core was written by American Airlines. We then returned to the Executive Office (EXO) to begin defining how the Eastern version of Sabre would be modified to fit United's needs, which were different and more intricate. We would make thousands of modifications to the Eastern System to accommodate United's requirements.

Also, at the same time, United's property development department began building a computer center in Denver and ordering the computers, peripherals, and storage devices (drums and disk drives) that would be needed. Programmers were being hired and trained in the ACP language. I picked up ACP in my spare time in case I needed to read what the programmer coded, for example, when checking on specific logic or computational complexity.

Our objective was to have the computer center built in six months and the entire system up and operating in twelve months. All that existed in Denver when the decision was made to build the computer center was a vacant lot. Excavators began digging the foundation without knowing exactly how big the building would be. They excavated deep enough for the computer room portion of the building to be below ground (protection against storms) and have good drainage. The construction was completed, the IBM equipment installed, and the programming well underway within the deadline.

One other challenge the company faced was what to name the system. Marketing initially called it Daedalus, after the Greek mythology character who engineered a set of wings for himself and his son, Icarus, to escape Sicily. But Icarus flew too close to the sun, at which point the wax holding the wings together melted, and he fell into the sea and drowned. Whoever thought this was a good name for an airline reservations system did not think it through. So, before the system went live, the name was changed more appropriately to Apollo. At least in Greek mythology, Apollo was one of the twelve Olympian gods and the son of Zeus and Leto. He was the god of many things, including music, poetry, prophecy, medicine, and archery. He was also associated with the sun and light and was sometimes referred to as the god of the sun. This was a good name for a system that is still running fifty years later under its subsequent name, Galileo.

Old to New System

Once the system came to life, Barbara Allen's team became testers. After testing, we became training and implementation coordinators for United's twelve Reservations Offices and ninety Ticket Offices. I was assigned to the New York, Washington DC, San Francisco, and Los Angeles reservation offices. We supervised the training on the new system and the data conversion from the old system. In addition, we arranged for the personnel from the Ticket Offices to be trained at the Reservations centers.

The system we helped design, develop, and test worked flawlessly -- Barbara Allen's team's diligence and hard work paid off. The implementation went very smoothly. However, some unexpected and interesting experiences occurred along the way.

My first cutover was at the New York Reservations office. When I say cutover, we would turn off the old way of making reservations and turn on the new one at a specific time, initiated by making a telephone call to the Denver Computer Center. The old way had two parts. The first was an electronic terminal that kept track of the available seats on each airplane for each departure. The second was a card, the size of an IBM data card, on which the passenger's name and contact information was handwritten. The card was then inserted into a conveyor belt slot which zipped the card to a backroom where keypunch operators keyed the data into a computer terminal for storage on a local computer. The operation also included a bank of teletype machines for communicating with other Reservations Offices and Airports and telephones to contact other airlines. In the early 1970s, airlines would take a booking on any airline and issue tickets for travel on other airlines, for when a connection required a change in carriers. Tickets from one airline were usable on another airline, if necessary to get to your destination.

In New York, our cutover was done late on a Sunday afternoon. Because we would be out of operation for a couple of hours, the staff of agents was reduced to a skeleton crew. To our shock, as soon as the new system started, it notified us that a Schedule Change had occurred. This meant that lists of passengers were inserted into an electronic Queue (a list in the computer) to be worked by agents. The agent had to call the customers to tell them their flight times had changed.

Because the staff was so thin, the manager, two supervisors, and I jumped on the phones and called all the passengers who had flights the next day to tell them their new departure times. We left the flights for the following days to be called by the dayshift on the next day. I got a first-hand taste of what the Reservation Agent's job was really like.

I then moved to our Washington, DC, Reservations Office. All the offices had two groups of Agents, one to handle the public and the other, the Mainliner team, to handle Travel Agents, who were more professional, experienced, and demanding. I had not had any problem with the Mainliners in the NY office, but DC was different. The General Res Agents were trained first in groups, and then the Mainliner agents were sent in two shifts so that Travel agent support would not stop. While the first half of the Mainliners were in training, the remaining agents were grousing loudly about how change was unnecessary and worthless.

I got quite an earful. Then the second group went off to be trained. When they returned, one Agent became overly critical and loud. First, she said that it was not going to work. Then, she loudly announced that she had proof that it was not working properly. I asked her what was not working. She said that she had put in a transaction code of RR, which meant that a reservation was requested from another airline. She complained that the system had changed the code to an HK, meaning the flight was confirmed, and she had not called the airline yet. I explained that that change was correct because Apollo had electronically contacted the other airline's computer and had instantly confirmed the request. To this, she responded loudly, "Wow! This is great!" Instant conversion of a loud complainer into an even louder advocate.

My next stop was the San Francisco Office. This was a large office and was managed by a friend of mine. He trusted me to manage the cutover on my own. The process went very smoothly until the time came to pull the switch on the Teletype Operators. They had a bank of twelve Teletype Operators working on a raised platform at one end of the Office. The Reservations Offices were large rooms with many small workstations for the Agents. However, in this office, the Teletype Operators worked in a separate open area in four rows of three, with all facing in one direction toward me.

I was standing in front of them using a telephone on a supporting column when the time came to instruct the Denver Computer Center to switch over the teletype function to the computers. I watched as all the teletype machines went silent. Then, in unison, all the operators looked up at me and said, "what happened to our machines?" In that instant, I realized the management had not told them their jobs had been eliminated. I also knew the Office Manager had already gone home for the day. I was the only management person left in the place.

Delivering Bad News

I apologized to them, explaining how the company was making this massive change in how our Reservation Offices operated. I also said that HR would find new jobs for them, which our team had arranged in planning this major conversion project. They all took the news very well. I remember feeling particularly sympathetic for a young African American gal, thinking she might find finding a new position more challenging than the others.

However, she was happy and outgoing, conversing with me after the others had left. Earlier, she had complimented me on the bright tie I was wearing (one of my trademarks around the office). She told me, "My brother would love that tie." We talked for a while, and I began feeling bad for her. Finally, I gave her my tie as a gift for her brother. As soon as I did, she thanked me for it and commented, "Oh, my brother will love this. He's the middle linebacker for the 49ers." I had just given my tie to an athlete who earned ten or twenty times what I made, if not more. I felt like a major dufus. Lesson learned: never assume!

Also, take nothing for granted. While managing the cutovers, we also set up a Central Reservation department for Apollo, where system experts would run a help desk, where schedule changes were entered, and other core functions of the system were managed.

As we were installing the new computers, the technicians could not get one terminal working. Technicians from AT&T, IBM, and the IT department surrounded this stubborn CRT. They were trying to figure out if it was a terminal problem, a software problem, or a communications circuit problem. No one could figure out the problem.

I leaned against a wall watching the dilemma unfold next to one of our construction men. Finally, the construction guy took something from his pocket, walked over to the wall where the terminal was plugged in, pulled out the plug, and stuck the voltmeter he was carrying into the wall socket. Sure enough, the meter read only eighty volts. A common man, an expert tradesman, showed up all the technology experts, a lesson never lost on me. Respect everyone because everybody has something to contribute.

FAMIS: Food Accounting and Information System

After cutting over the Los Angeles office, on Barbara Allen's recommendation, I was promoted to lead a team with the same composition and objectives as Barbara Allen's for the newly created Food Services division. This was my first job managing others formally.

You must learn to lead if you envision creating a large computer system or any other major project. How? By working your way up the chain of command and not screwing up on the way. This promotion was my first step onto the ladder of management. I was determined not to fail.

Creating and implementing large projects requires leadership skills. Leadership is a learnable skill[26] but takes time and experience. It means being willing to take responsibility at every opportunity and having the willingness to take risks. Risks can mean failure, but even a disaster can often be salvaged or turned around.

The old Arny motto, "Never Volunteer," was not a maxim I obeyed. Volunteering is how you broaden your knowledge and gain experience. Toss your hat in the ring for special duties, committees being formed, and task forces being assembled. Be the "go-to" person. Be the "jack of all trades." You cannot lead from the rear.

I discovered along the way that physically performing a task, especially considering my dyslexia, was more effective for me than reading about it in books. This is the reason why I never considered becoming a "professional student." For instance, common sense cannot be learned from books; one must go through the experience. However, as I mentioned, I did need to teach myself to speed-read technical manuals.

One other thought about creating and implementing big projects. Sometimes when you get an idea or set out to work on a project, it may not feel like a big project, but often once you get into it, the project grows. For a small-scale example, say you build a website. You might have to create a supporting database and then realize the need for encryption to protect the data. As you progress, you may even require maintenance screens for administrators in case of any mishaps. Then find you need a Privacy Policy, and so on. The same scope expansion occurs on large computer projects and must be planned for. Overall, my conclusion is that anything that creates value is not easy, or else everyone would do it.

As head of Applications Planning for the Food Division, I reported to the Division CFO, Anthony (Tony) Chaitin. He would later be promoted to SVP-Corporate Services.

Before this new division was formed, United operated Flight Kitchens at all the major airports. The kitchen managers reported to different local Station Managers, who mostly had no professional food service experience, leading to inconsistent spending on meals and quality control issues. This organizational structure resulted in cost-cutting measures that affected the quality of meals and left customers dissatisfied.

Minimizing costs was not the best goal for customer satisfaction or growing United's customer base. Therefore, food service needed to be standardized. Eddie Carlson, our CEO, understood that. Eddie was Chairman of the Board of the hotel company that United acquired, Westin Hotels. When he joined United's Board, he thought he had begun his retirement. However, he was soon called back into service when he was convinced to take on the Airline's President and

CEO role. During my time there, the airline cycled back and forth through phases of centralizing and decentralizing. Consolidating the Flight Kitchens was the latest centralization move.

Eddie Carlson chose one of his "hot shot" proteges from Westin Hotels to be President of the Food Services Division, Dick Ferris. When Richard Ferris arrived, he sent out a memo telling everyone that business was too informal and that everyone should be addressed by their last name, saying we were to call him Mr. Ferris. The next day, purely coincidentally, a memo was issued by Eddie Carlson, Ferris' boss, that the business environment at United was too formal and that everyone should be addressed by their first names. That contrast paints a picture of Dick Ferris' personality compared to Eddie Carlson's.

My boss, Anthony Chaitin, VP of Accounting, charged me with forming a team to create a fully integrated Food Accounting and Management Information System (FAMIS). Its purpose was to consolidate and integrate all the reporting by the Flight Kitchens and connect the results to the Corporate Accounting general ledger and payroll systems. Fortunately, I had a solid grounding in accounting practices from my MBA studies. The goal was also to standardize all the processes and food quality across the airline.

I received approval from Corporate IT to have the system run on the same computers as the Reservations System in Denver and be written in the ACP language. This allowed the company to utilize programmers coming off the large Reservations project as it was finished.

My Applications Planner consisted of three technical people and three procedure writers, comparable in the mix and number to my previous team. I remember the names of most of them: Jimmy Goodman, Lenny Sacchitello, and Charlie Primm were the procedure writers, and Reed Deemer, Larry Zerwas, and Sandy Miller were the technical Applications Planners. Carole Steinke was our Administrative Assistant. She was one of the fastest typists I ever encountered.

The Application Planners' job was to write the Functional Specifications spelling out the inputs, processing, and outputs. At the same time, the procedure writers, who had all worked in Flight Kitchens, would write out the operational processes surrounding the system functions and develop training programs and materials. It was a great crew with many foibles.

For example, with his great personality, Jimmy Goodman would craft presentations for me, except he would always unknowingly misspell a few words. He was notorious for that, and even when I pleaded with him to check and double-check the spelling in an important presentation, I still would catch at least one misspelling.

Charlie P. had worked at United Flight Kitchens for many years before I hired him into a headquarters job. He was a good fit, being both experienced and smart. One day I needed someone to visit the Miami Flight Kitchen, and I selected him for the mission because the trip required an experienced hand. He pleaded with me to send someone else and resisted my directive, I thought jokingly. Finally, I had to insist, and he reluctantly agreed. He left on a Thursday.

Come Monday, he did not show up in the office, so I asked the others on the team to find out where he was. They came back and told me he was not back from Miami yet. I found that strange because United had multiple daily non-stop flights there and back and the work would not even take one day. Then, finally, Carole explained that Charlie was afraid of flying and had driven there. How could a guy work for an airline and one that offers amazing travel benefits and not fly? I was shocked. I would have sent someone else if he had just told me.

Then there was Lenny Sacchitello. On a visit to the Newark, NJ Flight Kitchen, the kitchen manager escorted me around. When we got to the storeroom where all the dry supplies were kept, I was impressed with how neat and organized it was. I was then introduced to the Food Service Clerk, Lenny, who was the organizer of his domain.

I immediately knew that I wanted him on my team, so I made him an offer when I returned to the office. This was a multi-level promotion from a near-minimum wage position to a second-tier management position. It was a nice bump in pay but also included a paid move for his family from Newark to Chicago.

However, he was from a very tight-knit Italian family. So tight that he and his wife went to dinner every Sunday at either his mother's house or his wife's mother's house. I can imagine they all lived within a block or two of each other. He seemed a little lost and distracted when he first started. After a few weeks, I learned that he and his wife had been flying back and forth to Newark every Sunday for dinner, and his flight benefits were running out.

I counseled him several times about the promotion potential of working at headquarters and the importance of becoming more settled in Chicago. Each time, he said he would try.

A few months later, the first thing Monday morning, he came storming into my office ranting about wanting me to transfer him back to Newark to his old job. He was adamant. And I was shocked thinking he had settled in. When I asked him why, he said, "Last night, when my wife and I came home, there in our driveway was a rat this big," holding his hands about two feet apart. I responded, "Lenny, that was not a rat; that was a possum." He asked, "What's a possum?"

I explained, and from that day onward, he settled in. They had bought a house for the first time. Then, one weekday, when Lenny was on a business trip, his wife called me at home and asked me to help her because her water heater was leaking. She did not know whom to call. Of course, I was happy to assist. They were both still just kids.

And soon after, Lenny learned why you step on the beams when you walk in your attic, not on the sheetrock. He fell through straddling a beam. But, with his typical good luck landed with both feet on a shelf in a closet and was unharmed.

Lenny eventually retired from United Airlines as the Food Services Director of the Central Division, and our families remained friends. The others were all similarly interesting, smart, and talented.

The Design

The system required extensive design because it had to track what food would be provided and when as well as its food and labor costs. The menus varied by the kitchen, type of flight (short, medium, or long), and the class of service (First or Coach). The system had to adjust as reservations were made or canceled.

All the food ingredients had to be ordered and, when they arrived, tracked from their raw state to their finished form. For example, if a beef roast of ten pounds is bought, it might weigh 9.5 pounds after trimming the fat. Once it is cooked, it might weigh seven pounds. After being further trimmed and sliced, it might net six pounds or 96 ounces. It might then be portioned out to, say, thirty-two meals. If that roast costs $100, the meal cost would include $3.15 of meat. All the food items and costs had to be tracked from planning through ordering and delivering to the exact number of meals on flights, then charged back to the operating side of the airline. The accounting term for this is Yield Management. This is one example of complexity. There were many others.

Because we were now a separate division, the expectation was that we would charge the three Operating Divisions for the meals provided for their flights. Operations would have some menu input, but the kitchens would manage the rest.

A major debate within the airline, as well as within our division, became whether to charge the Operating Divisions for the meals using a fixed Standard Cost, which Operations would try to renegotiate all the time, or to charge like regular food businesses do -- using a 150% markup of the base food cost (divide the cost by 40%). The resulting price covers labor and overhead but places the burden of controlling costs, food, labor, and overhead on the kitchen.

For example, a coach meal with a food cost of, say, $3.50 would be charged to the division at $8.75, with the difference being applied by the Food Division to cover its labor, supplies, and overhead, and leave a typical food service profit of 20%. The alternative of using fixed Standard Costs might have said all Coach meals would be $7.00 or $9.00 without regard to the cost of the food in the meal. This approach would be simpler to program, but food costs could be out of control without anyone knowing, and having a separate profit center would become a farce. I suspected the Operating Divisions pushed for the standard cost approach because they secretly wanted our division to fail so they could regain control of the kitchens. A typical business political battle was brewing.

At one point, early in the design of FAMIS, I was called into our division president's office to discuss these two options. Dick Ferris pressured me to go the Standard Cost route, but I knew it would cause our division to fail and be reorganized, so I argued hard to keep to the Food Cost markup approach. With the support of my boss, Anthony Chaitin, Ferris finally acquiesced, saying as I walked out, "If you're wrong, you're fired!" Ouch! Tough love motivation?

Three months later, when we were heavily into finalizing the system's functional design and programming was underway, I was called into Ferris' office and forced to defend my decision again. Again, I had to argue hard to keep our direction, and this time my boss stayed silent on the issue, either because I was eloquent or because he was worried that I was wrong. I never did find that out. As I walked out of Ferris's office, Dick again said, "If you're wrong, I will fire you." I was too naive to ask what reward I would get if it worked. In hindsight, I wish I had.

When typed and stacked, the team wrote functional specifications that stood five feet high. We fed the specifications to the programming team in Denver as we finished them while working on more and while participating in testing and planning the cutover of the system.

At one point, the programmers were stuck trying to get a program to work. Everyone had looked at the code, and no one could find out why it was not working. I asked to see the code, which I usually would not do in my position. I studied the code for hours and could not find the problem. I hated it when failure was staring me in the face. I kept looking it over into the night. Then, at 3:00 AM, I suddenly saw the problem. Someone had typed the letter o in place of the number zero. The computer did not flag that as an error but indisputably misinterpreted it. When you must be hands-on, you had better have the necessary experience for what you are doing.

FAMIS Cutovers

When the system was ready and we were prepared to begin cutovers, we did a roll-out to the management team, which included all the chefs, at a conference in Portland, OR. The system was about to impose a great deal of discipline on the Flight Kitchen management team as the system was built to provide a daily Profit and Loss statement to the kitchen manager. I said DAILY! No one in any business in those days received an accurate P&L statement at the end of every day. They were lucky if they saw one two weeks after each month's end. The managers would see immediate results from their buying decisions, their supervision of their union workers – productivity and hours, and waste prevention. Hundreds of computations were needed to produce the daily P&L.

The conference included breakout sessions for the different kitchen supervisors, including one that included all the chefs. At one point, I was walking past the room where the chefs had just been introduced to the system when two of them stepped out, collared me, and dragged me into their room. I thought momentarily that it might be to beat me up. But instead, they wanted me to explain how the system would work for them.

After I finished explaining, I was surprised they were not mad but rather pleased because the system was well thought out, and they could see the benefits. They liked the immediate feedback aspects particularly. Thanks to my hard-working and intelligent team, the system rolled out smoothly and successfully.

My interactions with the chefs during my visits to flight kitchens often resulted in me being spoiled on subsequent trips. I remember visiting the San Francisco flight kitchen where world-class chef, John Wolfsheimer, was head chef. He always had his team prepare something special for me to eat, at any time of day. He liked to showcase their skills. This reminds me of another chef experience I had in San Francisco.

During an overnight stay at the St. Francis Westin Hotel downtown, I introduced myself to their head chef out of my general interest in food service. He kindly offered to show me around their kitchens, starting with the first-floor public restaurant kitchen and then the catering kitchen on the second floor, which featured a two-story oven capable of cooking hundreds of steaks at once. Finally, we arrived at the compact and busy kitchen of the top-floor fine-dining restaurant, where the head chef gestured toward two long workstations and declared, "These are our European-trained chefs." When I asked, "All of them?" He replied, with a wave of his hand, "No, just the ones at the first workstation." The chefs were segregated into two classes. I realized I had just learned a lesson about how chauvinistic a European-trained head chef can be.

Around this time, I acquired the company's first desktop computer. It was a Compaq portable computer, and we were the only division in the company to have one and I was the only one using it initially. I used it for designing the accounting reports for FAMIS and for doing our Divisional Budgets for my boss, the CFO. On it, I developed an early prototype program that performed the main functions that Microsoft's EXCEL would one day provide. My model used pseudo-English-language instructions and could add, subtract, multiply, divide, compute percentages, subtotal, total, add lines and dotted lines between sections of numbers, and format to a limited degree. I began to appreciate the future for desktop computing. I could see the potential of desktop computers from the beginning.

During the late 1960s through the 1980s, IBM's Selectric Typewriter competed in the typing business, offering greater control over font styles. It faced direct competition from Wang Word Processors, which excelled in advanced word processing capabilities, electronic editing, and document storage. The IBM Selectric typewriter was preferred for fast and reliable typing without extensive electronic document management. However, those needing advanced word processing functionalities leaned towards the superior option of Wang word processors. Word processors, including those from Wang Laboratories, revolutionized office workflows as technology advanced, eventually replacing typewriters as the dominant text-processing tool. Wang Industries, led by Dr. An Wang, was renowned for its word-processing machines and minicomputers.

While many typists used Wang Word Processors, the industry initially failed to foresee the disruptive potential of computers. Upon recognizing desktop computers' potential to outperform Wang's devices, I expected a swift pivot by Dr. Wang into programming on desktop computers. Regrettably, he did not. He might have been reluctant to acknowledge the risk of technology replacement, and his hardware-oriented company culture hindered a seamless shift to a software enterprise, necessitating significant operational and cultural adjustments. During these years, I learned both devices at a modest level of proficiency.

In 1978, I walked into a Radio Shack store for some electronic parts for my stereo. While waiting for help, I watched a young boy, age six, reaching up to type on a Tandy TRS-80 personal computer that sat on top of stacked boxes. He was typing away even though he had to stretch to reach the keyboard. It struck me that if a six-year-old relates to these new devices, I had better learn about them as soon as possible because it was the way of the future. It cost me about $800 (equal to $3,872 in 2024) to buy a TRS-80 and printer that day, and I never looked back. I have owned anywhere from one to three personal computers ever since.

Once FAMIS was fully implemented, as I predicted, the Operations Divisions came strongly after Dick Ferris about their new food invoices. They even called in corporate auditors to evaluate and rule on what we did. Because we had worked closely with the Corporate Accounting Department on the system's integration, they knew what we had designed and wholeheartedly endorsed our approach.

After my success in the face of Dick Ferris' threats, my boss awarded me one of the ball bearings he kept in the pencil tray in his desk drawer – his idea of an award for having the "balls" to stick to my guns on the pricing approach. That was the only recognition we received. It amazes me how often major accomplishments in business are not recognized.

During this project, I regularly wore loud neckties, as my "trademark," and partly in rebellion against Ferris's original formality and maybe authority in general. Every time he walked past my office and looked in and saw my loud tie, he would grumble and shake his head. He disliked my neckties – it was not a joke to him, but it was to me. He disliked me for my independence but had to tolerate me because his success depended on what we were doing.

After our successful cutover, Ferris organized an Annual Division Meeting. He arranged the room so he and his VPs would sit at a long table on a dais at the front, looking down on the rest of his management team members (like serfs). The room had about twenty rows of chairs, with seven people on each side of the center aisle. I intentionally took a right-hand seat on the aisle about halfway between the front and back. We all stood up as he came in and walked down the aisle, like he was royalty.

When he came to where I stood, he stopped and attempted to embarrass me by mocking my loud tie. I was ready for him.

I said, "Dick, you have not seen anything yet," and opened my suit jacket, allowing my huge tie to unroll down to the floor. It was about a foot wide at the bottom. The room erupted in laughter. He was utterly flummoxed and quickly proceeded to his chair on the dais and started the meeting. He never mentioned my ties again.

One would think my impertinence would be political suicide in a big company, but Ferris was destined for greatness as Eddie Carlson's fair-haired boy. Ferris soon left to become Group Vice President of Marketing, and then, when Eddie Carlson moved up to Chairman, Ferris became CEO of the airline in 1976. He also became the Chairman of the Board and remained CEO until 1987. I rarely saw him after he left Food Services. His parting speech when he left Food Services included a statement that he would not forget us. But he did. He would go on to drive the deregulation of the airlines. If he were still living, you could thank him for the resulting low airfares and no legroom.

Ferris never thanked me for the risks I took or the quality work my team did for the system's success. However, his VP of Operations did thank our team with a brief celebration at the Denver Kitchen! And, of course, our team celebrated among ourselves.

Later, my boss, when he became SVP-Corporate Services, offered me the position of Director of Payroll, which I declined. I did not see a career ladder leading upwards from there, and I found limited opportunity to be creative in accounting, although I had learned a lot about accounting at the University of Chicago and while creating FAMIS

.

My New Boss and Super Leader

Jim Kent replaced Ferris. He was the best delegator I ever met and one of the most extraordinary leaders I knew. If you walked into his office, you would be shocked to see nothing on his desk, and I mean nothing. Any paper he needed to sign would be brought to him, he would read it (or not, if he knew what it was) and sign it at once, and it would leave with the person who brought it. I once asked him why he worked that way. He answered that he was paid to think, not shuffle paper. He trusted his direct reports to do their homework and manage the business details.

At one point, Jim had begun organizing a management conference and had assigned a director to manage it, but the director was not getting the job done. So, he asked me to "go to work for the director temporarily" for this project but, in private, asked me to take over the organizing and get it done. He left the director "in charge" but told him to give me free rein. This was a bit delicate, but Jim knew he could trust me to get the job done.

I pulled in Jimmy Goodman to help me because there was much to be done quickly, and Jimmy was a doer. We did it "in spades" and documented every step and process as we discovered them.

The theme was "Excellence." We plastered that "conference brand" on everything, including the paper strips that the hotel wrapped around the toilet seats after they cleaned them. We had bellmen place engraved invitations sealed with sealing wax on each bed the day before every session and much more. Our chefs planned all the meals. The event was an enormous success.

Jim Kent gave all the public accolades to the director but complemented Jimmy and me privately after the conference concluded. It was the most organized conference ever.

Soon after, Jim Kent needed an Assistant to handle some special assignments -- he already had an Administrative Assistant. Fortunately, he offered me the job. Unfortunately, I could not bring Jimmy Goodman with me this time, but he was always nearby to help when I needed him and would later have a remarkable career.

One of the first projects I was asked to address as Jim's Assistant was the creation of a "Management Information System." He kept asking his VPs for this System in our weekly staff meetings. His VPs were as confused as me because we had just created FAMIS, a Management Information System, but he was asking for something else. Jim kept asking his VPs to the point where everyone was frustrated. Finally, the VPs came to me for help. Later that day, I walked into Jim's office and explained to him how no one understood him and that I needed him to explain it to me so that I could explain it to the others. In that discussion, I got a clearer understanding of what he wanted.

He wanted an elaborate system, on paper, which included documenting all company **policies, procedures, standards, guidelines, and checklists**. This was a monumental undertaking because these five categories of information appear at every organizational level (e.g., company policies, division policies, department policies, personnel policies at multiple levels, etc.) The resulting documentation ended up filling entire credenzas with large three-ring binders. I not only had to educate his VPs and create definitions and formats, but I also had to educate the entire Division.

However, the value was not in the resulting documentation (i.e., the destination) but in assembling all the information, finding missing information, writing up all the missing details, and getting approval for previously undocumented policies (i.e., the journey). Everyone had to identify those five classes of information for their position or level in the business, understand them, and sign off on all those documented below their place in the organization. This was a massive exercise in communicating, discussing, understanding, debating, and defining. The outcome of this extensive exercise resulted in the best-organized operation I have ever seen.

Jim had been the Vice President de Mexico for Westin Hotels and had used his system throughout his career. He believed in being highly organized and would tolerate nothing less. He once was assigned to head up a hotel in Kansas City. Over the weekend, before he was expected to arrive, he went to the hotel and with the help of the bellmen, removed everything from all the offices of his direct reports.

When they arrived on Monday morning to find their offices empty, panic ensued as they feared they had lost their jobs. Jim called them together and asked them if all the information removed from their offices was important to do their jobs. When they acknowledged that all that information was important, he instructed them to keep it organized going forward, according to his system.

Years later, I arrived at a Houston, TX, hotel for a conference. I had never heard of Jim having anything to do with this hotel. However, just walking through the hotel, I could tell he had worked there. I asked five or six hotel employees if they knew or had heard of Jim, and none recalled him being there. Finally, I asked an elderly waiter during the conference dinner if Jim had ever worked there, and sure enough, he had. The waiter remembered Jim running the hotel years before. I call that "having an impact" when I could feel his presence 15 years later. Within Jim's Management Information System, I was assigned the job of creating the Personnel subsystem with assistance from a Ph.D. personnel consultant.

The subsystem included a Career Ladder, Job Description standards, Annual Review Processes, and training objectives to assist the person holding the position in preparing for advancement. The Job Description Standards included the desired skills and experience for someone to fill the position. In addition, the subsystem had whether the person was ready for advancement or, if not, when they would be or if never, why not.

After implementation, employee satisfaction improved significantly. This was attributed to the clear understanding employees had of their positions, future opportunities, and how to prepare for their desired roles. Additionally, the system provided transparency regarding the timing of their next promotion, or if they were at a plateau in their career progression. Interestingly, those few classified as plateaued did not mind, already knew where they stood, and appreciated knowing how they were perceived. A person can plateau and still be respected as a professional. It would just be that their compensation increases would be limited to inflationary adjustments.

Early on, Jim Kent asked me to arrange a whirlwind tour to inspect all eleven Flight Kitchens. I did not travel with him on this trip, but I let every kitchen manager know what to expect before he arrived. The kitchen manager at O'Hare was a particularly enthusiastic manager who wanted to ensure his kitchen would be ready for inspection. I warned him that he would not likely outsmart Jim Kent, but he said they would be prepared.

As I heard later from the manager that when Jim arrived, he spotted that the kitchen was in top shape, so he asked the manager to see his roof. Roof? The manager had not even seen the roof himself. As soon as they got on the roof, Jim pointed out several things that needed to be done. The manager was embarrassed. Jim then walked the manager to the roof's edge and pointed to the lot next door littered with paper. Exasperated, the manager explained to Jim, "But that is not our property." To which Jim responded, "No, but it is your paper. Clean it up." Every inspection went like that. Jim set the highest standards.

Temporary in Detroit

At another time, Jim suspected that the manager of the Detroit Flight Kitchen was out looking for another job, as he had extended his vacation by an extra week. The second in command was the head chef, Jerry Smith, the only chef in our system that was not European-trained. Jim did not trust Jerry to fill in for the manager for more than the first week. Besides, Jerry was needed in the day-to-day operation. So, with Jim's approval, the VP of Operations asked me to go up and run the kitchen for a week. As it sounded fun, I agreed, wondering if I would be offered the position if the vacationing manager decided not to return.

When I arrived and met the Head Chef, I found him loud, brusque, and insensitive. I also learned what running a 24-hour-per-day, 3-shift, seven-days-per-week operation was like. The manager needs to know when to go home. I spent 18+ hours every day because there was never a break in the action.

I made a few changes while I was there. I met a couple of inbound flights and asked the flight attendants what they might need on their outbound flights that we were not providing. One item that the airline had provided for a brief time was ice cream sundaes in First Class. The attendants told they had been very popular, so I reintroduced them. The next day I added root beer so that they could offer root beer floats. After putting both on the flights for a few days, I got a call from the Midwest Operations VP (of the airline), complaining that I was increasing his meal costs because, remember, our Food Division calculated his charges based on marking up the underlying food cost. So, it was his call, and I had to stop that. I was a hero for a couple of days with the Flight Attendants and some First-Class customers.

I found the Chef amusing, cagey, and adept at handling the unionized kitchen workers. One day, when we needed to catch up on getting the meals onboard, I asked the chef how he could help. He said, "Oh, watch this." He then entered the cooking area and began cooking, a task reserved for a union cook per the union contract. When the cooks complained, he hollered, "I'm training you. Pay attention." By union rules, he was allowed to cook while training them. The union cooks shook their heads, acknowledging that their Chef had outsmarted them.

I got a copy of the union contract, a small book that would fit in your back pocket and read it through twice. I learned that having an intimate knowledge of the contract was useful when dealing with the shop stewards. This experience prepared me for what would follow later.

Another time, we were getting stacked up on the loading dock with dirty buffets, the metal cabinets that roll onto and off airplanes. Looking into the situation with the Chef, it was clear that the backup was concentrated around the front end of the large dishwashing machine, close to the loading dock. The dishwasher was a long conveyor belt surrounded by metal walls and pipes spraying hot water on the dishes. The dirty dishes were placed on one end, and clean dishes came out on the other. Obviously, we could not move the dishwasher.

The Chef and I studied the situation and could not see a solution. The Chef went back to the cooking area while I remained. Then I got this brilliant idea – why not ask the guy who ran the dishwasher what he thought? He was shocked when I asked, saying, "No one ever asked for my opinion before." To solve our problem, he suggested we should line up the buffets, starting on the clean end, where there was more room, and then he would move them toward him as he worked each one. He would push them out to the empty side of the loading dock when cleaned. The problem was solved simply by asking the guy who did the job day in and day out. Another lesson learned!

When my week was up, I learned that the manager would be returning on Monday, so any chance of me being offered a permanent position there was now moot. As my last act, I sat down with the Chef and counseled him on his behavior, which, as I mentioned above, was very blunt and insensitive. Nevertheless, he seemed genuinely concerned and wanted to know what he could do. I suggested he take a Dale Carnegie course named after Dale's book, How to Win Friends and Influence People.

Six months later, Chef Smith stopped at our headquarters office. I asked him how he was doing, and he said he was doing better than ever. He not only took the course I recommended but told me that he was now teaching it. I checked with our corporate head chef, who confirmed that the man had completely changed his personality and was doing great in his job.

Always Be Ready for Change

One last story about Jim Kent. One day during our Monday weekly staff meeting, he asked his VPs, Administrative Assistant, and me if our calendars were clear or could be cleared for the afternoon. All agreed they were or would be cleared. He asked us to look out of the window. When we did, we saw a stretch limo waiting outside that his wife had brought for us. He would not tell us anymore. We all went down and got into the limo, and it drove off toward Chicago. He had been in the division for a couple of years, so some of us wondered if he was about to announce that he was moving to a new job.

When the limo pulled up in front of the Chicago Art Institute, he told us we were there to see the Monet Exhibit. I had known this famous exhibit was happening and wanted to see it, but I had not taken the time to do so.

To describe what we saw, Monet painted in three different stages in his career. In his first stage, he painted accurately -- still lifes and portraits. In his second stage, he painted the impressionistic scenes for which he became well-known. Finally, in his third stage, he painted huge canvases of water lilies where he tried to perfect paintings with depth, but no horizon, which was the primary way depth was painted in his day. We all found this visit exciting but could not understand Jim's purpose yet.

We then returned to the limo, which drove us to the Cape Cod Room at the Drake Hotel for lunch. We had a great seafood lunch but were anxious to learn what was up. Finally, Jim said, "You probably are wondering why I invited you downtown today to see the Monet exhibit and have lunch. Well, I wanted you to realize that Monet had three separate careers during his life and that you should be prepared to do the same. You will have other careers, maybe in other industries, and I hope you will use what you have learned in your present career and apply it to those other careers."

That was it. That was his poignant message, which I still remember and will never forget. His prediction was true for me. The organizational skills I learned from him were applied to every job I had after that. He gave me the finest picture of how an executive should manage a large organization.

Contentious Negotiations

I was supposed to stay in the Assistant to the President job for one year, but Jim kept me there for three. A duty I inherited in the job was to sit on the Company's Union Negotiating Committee. The IAMAW (International Association of Machinists and Aerospace Workers) Union included all the ground personnel and mechanics as well as most kitchen workers. The cooks, pantry workers, kitchen workers, and food service clerks were all union, meaning that their work was limited to the duties defined explicitly in their union contract. You could not summarily fire anyone in these positions without following a formal grievance process and holding a hearing with an independent arbitrator. Since other union members significantly outnumbered the food service members, their concerns received little attention during contract negotiations. The Union officials' main concern was to keep mechanics' pay rates higher than the ground personnel and the food workers.

A month before the negotiating meetings began, I met with the kitchen managers to strategize what I might present as company positions in the negotiations. To identify what changes to the union rules might be successful, I planned a strategy based on the three constituents: the workers, the negotiators, and the company. I surveyed the managers to gather ideas that would meet the following criteria and be:

1. Acceptable to the food service workers.
2. Save money for our division.
3. Not particularly of concern to mechanics.

As a team, we first made a list of about thirty ideas and then boiled it down based on these criteria into eight changes that I then inserted into the Opening Letter to the negotiations.

The discussions went on for months. My role was as an advisor -- to listen and only talk to the Company lead negotiator privately outside the room. This is the nature of union negotiations. The union reps would complain, bluster, and posture for months. Then, when the old contract was about to expire, the lead Union rep and the lead Company rep (an attorney by trade) would go off into a private room and "horse trade." After resolving major concerns for mechanics, such as pay raises, medical benefits, and vacation time, the union would typically concede a few other requests with little significance to mechanics.

When the changes were deemed acceptable to the Union rep and the Company rep, we learned what the changes would be. Next, the Union and Company lawyers would revise the contract language for final approval. Then, as the last step, the members would vote on the contract, and the leadership would sign it if it passed.

The rank and file were primarily concerned with receiving regular pay raises and often overlooked the details of the agreement until they affected their work environment. If the members rejected the agreement, the committee would need to resume negotiations, or the union might call for a strike. However, we did not reach that point during the only negotiation I participated in.

After the negotiations were completed, I learned that the Union had agreed to most of our "lesser" items. Five of our proposed eight changes were accepted by the Union, including lowering some work from Pantry Workers to Kitchen Workers. By lowering work, I mean that work done previously by higher-paid Pantry Workers could now be done by lower-paid Kitchen Workers. Lowering work is unheard of in union negotiations except for this one time. This modification gave the Kitchen Manager additional flexibility to delegate work to other kitchen staff and promoted better teamwork between the two job groups. For instance, during the preparation of salads, which did not demand any specific expertise, now any kitchen staff member could assist.

Although some Pantry workers initially objected to losing some of their responsibilities to the Kitchen Workers when the modification was introduced, they eventually adjusted to the new arrangement, as we did not misuse the privilege.

After the contract was signed, both lead negotiators faced heavy criticism as they were accused of giving away too much. The deal proved to be a 'No Win' situation for them.

A Devastating Strike

The following year, the airline was struck by the pilot's union resulting in the shutdown of operations. I was not on the negotiating team for that contract. A large portion of the employees were furloughed while the remaining workforce continued to work in crucial positions at half their regular pay. For those who were laid off, this meant no source of income to pay for their mortgages and other bills, posing severe financial hardship. For those who continued to work at half pay, the situation was slightly better, but it still put their households at risk of foreclosure and financial hardship. The strike brought about a drastic situation for everyone involved.

I was retained at half pay and assigned to supervise a crew of about twelve fellow managers to keep the Executive Offices clean, which meant emptying the trash, polishing the floors, and cleaning the bathrooms. I looked at this "burden" as an opportunity to prove my leadership skills, so I assigned the easier jobs to the twelve managers and kept the worst job, cleaning the toilets, for myself. I did not simply supervise; I shared in the work. The job I kept was the worst of the jobs. But it did allow for partial pay.

However, my wife and I had to "amortize" a couple of mortgage payments, allowing us to defer payments for two months or longer as required. However, the principal and interest were rolled into the end of the mortgage with interest, leading to an increase in our loan balance and future interest payments. It was a setback financially, but necessary to manage our personal cash flow and keep my job for its prospects.

A Special CEO Up Close

From the time I went to work for Jim Kent, I became a student of what makes Senior Management tick, especially presidents and CEOs. I was interested in the skills that qualified them for that level. Many executives below the CEO seemed to be intelligent and have the potential for the top job. So, how did the top dog emerge from the pack? Spoiler alert: Emergence begins with who you know and, most importantly, follows with how much those above you respect and trust you, which only happens over time, if you are mistake-free. This is why many senior managers are risk averse. Those who make mistakes rarely are the ones promoted.

Eddie Carlson was a splendid example of a good executive. As I mentioned earlier, he retired and took a seat on United's Board of Directors when his company, Westin Hotels, was acquired by United. After attending monthly board meetings for about a year, the board coached him out of retirement to take over the airline as its CEO. He served as CEO from 1971 to 1979. In his first two years, he turned our nearly bankrupt airline into a profitable one.[15]

I spoke with him once while escorting him from the airport to our Portland Food Services conference, where he was scheduled to be our keynote speaker. Jim Kent asked me to drive to the airport to pick up Eddie and bring him to the conference hotel. I took Jimmy Goodman with me. The idea was for Jimmy to drive our rental car on the way back while my job was to listen intently to everything Eddie said. Jim Kent had told me to memorize everything Eddie said and repeat it to him word for word as soon as we returned. Remember, memorizing was not my strong suit, and I knew Jimmy would not be able to help much by being focused on the road. But I did my best. I heard nothing I recognized as earth-shattering on our drive. However, when I later recounted the conversation to Jim, he found meaning in what I related. I did not grasp the significance of what Eddie said, and Jim never explained it to me even though I asked twice.

One skill of Eddie Carlson that I immediately recognized was his ability to talk with anyone about anything. He was pleasant and friendly with everyone he spoke with. This is a learnable skill.

I also remember a story that Eddie told during that car ride. A few months earlier, not long after he accepted the position of CEO, he had left the company's Elk Grove Village, IL, office late in the day to catch a flight out of O'Hare. He had not eaten all day, so, he asked his driver to stop at the flight kitchen so he would not need to go into the terminal, and because he knew the kitchen always had food available for their kitchen workers.

In the employee kitchen area, he walked through the food line with his tray like any other worker, picking up the surplus food items provided by the kitchen from the meals produced during the day, like chicken wings. Like any kitchen worker, he took his tray and sat with three kitchen workers at a table. He asked each of them about their job and how long they had been working for United. Each talked about their title, work, and the years they had been with the company. They had all worked for the company for ten years or more. Then, one Pantry Worker asked Eddie his position and how long he had worked for United. Eddie answered, "Well, my title is President and CEO, and I've worked for United for about three weeks." To this, the Pantry Worker responded, "Wow! You move fast!"

This story portrays the kind of leader Eddie was. He always met with and asked what was on the employees' minds. He carried a small leather sleeve that his staff called a "Ready Eddie" containing 3x5 cards. He made a note on a card when an employee had a complaint or question that he could not answer. Then, when he returned to the office, he would dole out those cards like a Las Vegas dealer to his subordinates. Many grassroots problems were solved this way.

At this point, you might ask, what does all this have to do with an I.T. Career? First, these experiences began to teach me how senior management functions and how they relate to those they lead. Beyond knowing how to perfect the processes for designing, developing, and implementing large computer projects and I.T. systems like Apollo and FAMIS, I needed to learn about managing large teams.

Here is some of what I learned: Envision creating a large computer system project and how you would learn to lead the teams who will build it. No one completes any large project alone; many others are needed. Those others must be organized, know their parts in the process, and share in the success. Yes, there is a process for building a major system, but unique people skills are also needed in steering the process. Steering is a more accurate term for managing an organization than leading it. Someone must navigate and steer the boat, but everyone in it must do their part to propel the boat toward its destination. The boat's pilot is focused on looking ahead and relies on the team behind him to do their jobs well. Furthermore, progressing through an organization and earning experience along the way is valuable, often referred to as 'earning your stripes'.

From Jim Kent and Eddie Carlson, I was beginning to learn how to organize and motivate teams, negotiate business deals, relate to unions, and navigate upper management's politics. But there was still more to learn.

Operating Performance Committee & Big Data

Big Data existed before the term was coined, and I inherited a boatload of it in my next job.

When I finally left Jim Kent, not long before he left United to resettle in Canada, I moved into a new role, reporting to the Director of Operations Research, Herb Hubbard. The job had operational analytic responsibilities but also came with an additional title and reporting relationship.

I was appointed as the Chairman of the Operating Performance Team and reported directly to Percy Wood, the airline president at that time. Percy Wood later became famous for receiving a mail bomb from the Unabomber. He lost the ends of a couple of his fingers. It could have been much worse because Percy Wood was not wearing his glasses and held the package away from his body to read the name of the sender. The attacker was a lunatic environmentalist who specifically targeted him because his last name was Wood.

In this role, I led a team that was the only standing committee in the airline and consisted of multi-discipline members, each representing one of the airline's operating divisions. This standing committee's job was very broad -- to improve the operations of the airline.

This goal sounded simple enough, but an airline is a significant entity, including Aircraft maintenance with their hangers at big airports, flight and inflight crew management across the airline, flight schedules, flight kitchens, airport ground crew, terminal operations, freight, charters, and more. The scope was almost the entire airline, omitting only "overhead departments," like accounting and HR.

The committee consisted of staff-level managers and directors, intelligent people with no authority to change anything directly. You might consider them as Influencers. Our power came from getting the president to approve our recommendations.

Our first efforts were aimed at improving on-time performance, which needed to be a lot better. We developed a major new system that tracked every time an aircraft moved, took off, landed, or returned to a gate. The system also tracked every delay, its cause, and the delay's length. The amount of data was massive. The system then turned that raw (Big) data into performance reports for station managers, crew managers, ground managers, etc. We made it so that everyone in the operating management chain of the airline would suddenly have their performance measured daily against goals to improve from their past. No one wanted what we were building except the president because they felt their jobs threatened.

We moved the measurement standard from "on-time within five minutes" to "on-time zero." This set the expectation of zero tolerance for delays. In fairness, along with the change in standard, we lowered the goal. The goal was reduced from 85% on-time within five minutes to 65% on-time zero. We calculated that this change in goal would improve overall on-time performance by five percentage points, which it did. It was based on statistical distribution curves and standard deviations – in essence what we did was to move the curve. Fortunately, the goal reduction helped us sell this change to the managers because they sounded doable to the affected managers. Along with this change, we set goals for the average length of delays for the station and by department. This way either flights left on time, or when a delay was unavoidable, the responsible department would be focused on keeping the delay as short as possible. In a sense, we had delays surrounded,

We also identified the flights most likely to mess up the airline, each aircraft's first flight of the day. We then designated these flights as STAR flights for Start the Airline Right. This program provided separate tracking on those flights with more stringent goals.

First High-Level Language

As part of my regular job, I also developed an operating performance reporting system using a programming tool called FOCUS. FOCUS was created by Information Builders Inc., a software company based in New York City. FOCUS, which stands for "Fourth Generation Computer Systems", is a programming language and development environment. FOCUS was a unique programming tool because it interpreted English language phrases into computer code to process the information in the extensive database we had developed. FOCUS was absorbed into Tibco Software in 2020 and may no longer exist in its prior form. It is another example of the constant evolution of programming.

Unfortunately, FOCUS was one of those marvelous Big Ideas that did not quite catch on as it should have. Its main drawback was that its database design was hierarchical in structure, a method to which few programmers related. (Note: I will describe this type of structure later when describing our work with an Oracle database at Orange Lake.)

I had to program FOCUS myself because the regular programming department was too busy and resisted change. In fact, initially, they fought me over it. There was significant resistance to allowing me to install it on the company's main business computers, where my data resided. To overcome this obstacle, I had to navigate the political landscape and leverage my position as Committee Chairman, directly reporting to the president. I soon obtained the necessary permission to proceed. Although I went over the heads of certain individuals, which is a large-organization no-no, it was the right thing to do for the airline.

One day, after I got it installed and working, I ran into two of the IT managers in the cafeteria, both of whom had fought me on installing it. I talked them into coming to my desk and allowing me to demonstrate what it could do.

IT folks are always eager to learn about new technology when they are not trying to protect their political turf. As I showed them what it could do, I heard "Wow!" and "That's cool!" from the two managers looking over my shoulder. They never learned or used it for themselves, perhaps lacking the time to, not grasping hierarchical database structures, or not seeing how its capability might apply to other existing or new applications.

Using FOCUS, I produced for the president a monthly "pocket piece" that showed in charts and graphs all the pertinent operating performance compared to the past, trends, and the goals we set. Once programmed, I could produce the entire booklet of charts and graphs in minutes as soon as the data from the last day of the month was loaded. He loved it and would write questions on it occasionally and send it back to me for explanations, investigations, or a time or two, early on, a few corrections.

Denver Performance Problems

I also used FOCUS to solve a problem brought to me by Steve Grossveldt, United's Station Manager for Denver airport. His operating performance was the worst of all United airports in the country, and he was getting heat from the company president to fix it. He implored me to help him figure out what the problem was.

I flew to Denver to meet with him. As soon as I landed, I saw immediately what the problem was. Almost all the United Airline's flights into and out of Denver were using DC8s. Back in my office, for other reasons, I had already analyzed operating performance by aircraft type, and I knew the DC8s were notoriously the worst-performing aircraft, mechanically. Steve was immediately relieved once I explained his built-in disadvantage and gave him an easy explanation for his boss.

While on-site, I also investigated why the average length of his delays was higher than other station managers. Observing several flights, I quickly saw the cause of the problem. The DC8s were long-bodied and carried up to 189 passengers. It unloaded passengers from the front door using only one jetway. So, unloading the plane took a long time.

While the passengers were disembarking, the chief mechanic would wait for all the passengers to get off, taking about 15 minutes, and then go to the cockpit to get the logbook so he could check for any needed repairs logged by the crew, which was often the case. Then, if anything urgent needed to be fixed, he would have to order parts, wait for them to be driven over from the hanger, and install them. This meant at least another 20 minutes delay on average after the passengers were off. That meant 35 minutes were needed to check on many flights that were only allowed 20 minutes as "through flights." A "through flight" continues in the same direction it was headed, not returning immediately to where it came from, and gets minimal servicing, like tidying up. A "turn flight" must be refueled, have its trash emptied, have seat pockets cleaned, etc., before returning to its origin, a thorough cleaning.

I proposed an easy solution. The chief mechanic would use a pole, which he had on hand already but was not using and was made for this purpose. He would hold the pole up to the cockpit window, which opened, and the pilot could hang the logbook on the top end of the pole. This straightforward manual procedure trimmed 15 minutes off every delay. With these two changes accomplished in a few minutes of observation, the station manager was out of the president's proverbial doghouse.

Follow the Tail Numbers

Also, in this job, I discovered something few people in the airline knew. Flights were scheduled by airplane based on the plane's tail number. Each plane has a unique "FAA tail number," painted on the rear side of each airplane. All planes reside overnight at a specific airport where significant repairs can be done if needed. So, in a sense, all airplanes go "home" at night to a base that can repair that type of aircraft.

The department that built flight schedules did so by starting that Tail Number from its home airport and routing it through a series of airports, returning to its home at the end of its day. Each stop was allowed the number of minutes based on where it was a through flight or a turn flight, as explained above. In some airports, like O'Hare, extra time might be allowed for passengers from connecting flights to move between flights.

Once I learned how the scheduling worked, I assembled a database of operating performance by tail number and then wrote a program to process and report that data. In this case, the operating performance included the average number of delays per month, and the average delay length, by cause (mechanical, crew, check-in agents, etc.) By tracking these measures by tail number, I could determine precisely where delays were occurring along the route. I then worked with the station manager if the causes were controllable at his airport or worked with the scheduler if they had not allowed enough time in the schedule for flights through congested airports. This report fixed many causes of poor operating performance through this unique report that United had never had before. Again, the FOCUS language was critical in creating this report. I think that report died when Percy Wood and I left the airline.

Senior Management Presentations

As the Operating Performance Committee Chairman, I also made monthly presentations to the Senior Management Committee, consisting of the president and his direct reports. This was before the days of PowerPoint, so every presentation was produced with my content and direction by the art department on 35mm slides, and it took ten to fourteen days of lead time to do the analyses and prepare slides. No changes could be made at the last minute. If some bit of data or a chart were wrong, I would have to "talk in" the correction, which was unacceptable at that level.

I was so nervous presenting the first time that, afterwards, I was sent by the president to the head of the Inflight Training Center to be schooled in presentation speaking before I was allowed to present again. It was a perfunctory exercise because it was clear to the head of training in our first meeting that I had no problem speaking in public. The problem was my nerves and feeling like I was walking on unstable ground because the entire environment was new. And each presentation pointed out problems with how the airline was being run, with every problem being the fault of one of the executives in the room. And my current job and any future job I might get would be determined by one of the executives in the room.

By the third month, I became comfortable and proficient by learning a skill my boss gave me. The trick was "no surprises" for those in the room. If my team found a problem, we would figure out the solution, hand the solution to the Division Head responsible before the meeting, present the problem in the meeting, and let him explain the solution. That kept everyone happy and still allowed us to move forward as we saw the need. And it gave our committee much power.

I do remember one presentation where I found a severe performance problem and then, after the presentation, realized that instead of running the data using scheduled flights, I had used non-scheduled flights -- freight, charter, and other miscellaneous flights. It put me into a panic until I reran the report with the right flights. Fortunately, when I ran the correct data, the conclusions came out the same, and I did not have to go back and confess to a significant mistake that might have embarrassed me and a Senior VP or two.

In the end, with my team, the presentations became so productive that instead of presenting monthly, I presented every two weeks. In addition, the lead time on slide production meant I had to work out ahead of issues. This was one of the highest-pressure jobs I ever had or ever would.

My boss, the Director of Operations Research, Herb Hubbard, was quiet but brilliant. When tasked with explaining the economic impact of the airline's shift from circling approaches to airports to a technique called High Profile Descent, in three days he delivered a comprehensive analysis. The costs of the change included wear and tear on the airplanes, changes in the maintenance of the aircraft, impacts on flight and in-flight crews, fuel costs, noise abatement, scheduling changes, and several other factors. In short order, he presented a concise one-page chart that effectively captured the complete economic impact of the transition. He was a genius at analyzing and communicating complex concepts to Senior Management.

He also taught me an easy way to improve my writing. It was to look for phrases that included the word "of" and move the word that followed the "of" to be before the preceding word and take the "of" out. For example, if I wrote, "you have a pile of paper," I should have written, "You have a paper pile." Simple but effective.

How Safe is Flying?

I was personally curious about how safe it is to fly. Since I had all the data in my database, I did an analysis. I took all the "accidents," from denting a wing by bumping another aircraft on taxiing to the very rare instance of a fatal or non-fatal crash and "incidents", like running off a runway. I used twenty years of data to calculate the likelihood of being in an accident or incident.

The data showed that if you flew every day of your life from age 1 to age 80, the odds of you being in an aircraft accident or experiencing an incident was one chance in 300.

This means that the odds of being in an accident are infinitesimally smaller. I also attended many retirement parties for pilots who had flown for up to 45 years without injury or accident. I concluded that flying was very safe and is even much safer today, with numerous additional safety features added since 1978 when I ran these numbers.

Starting Flights to Mexico

I was given a special assignment to coordinate an FAA Check Flight to demonstrate the airline's readiness to begin service to Merida, Cancun, and Cozumel on Mexico's Yucatan peninsula. The flight was to operate in two weeks from the time I was assigned this task. On the surface, this sounded like a simple enough organizing event until I got into the details.

First, our airline had no operations in Mexico -- no ticket counters, check-in, ground-ground handling or fueling ability, or anything else in Mexico. We had no collateral or paper items, like ticket stock, boarding passes, or printed schedules, and, what's more, we were not allowed to import them into Mexico from the United States. And they needed to be bilingual. Printing them in Mexico was infeasible, given design and branding considerations, even if we had more than the two weeks we were given.

I flew down on Aero Mexico via Mexico City to learn the operating environment. As an introduction to flights operating in Mexico, the aircraft I was on landed so hard it knocked all the oxygen masks down.

I soon learned that most of the ground operations in the Yucatan were handled by one businessman. I met him and quickly solved most of the local challenges with his help. But I would need to bring all the paper collateral products with me, even though it would not be legal. The local staff would not be automated due to the lack of telecommunications capability from the U.S., so all processing of tickets and flight check-ins would be handled manually. So, they would need documented procedures and training.

He also said he would arrange a local reception event for the first flight's arrival at each location and dinner in Cancun with local hotel owners. Our plan for the demonstration flight was to land briefly in Merida, then fly on to Cancun for an overnight. We arranged hotel rooms at the hotel where the reception would be held. The next morning, we would fly to Cozumel for breakfast before returning home.

Second, I had to reserve an aircraft of the type planned for that market. I also had to work out the scheduled times for a flight on this path and load that into our reservations system so that I could book people onto the flight. Then, I reserved the first-class cabin seats for the president, the pertinent SVPs, whose departments were involved in the new service, and FAA officials and made sure that the food would be top-notch.

Third, we needed personnel in Mexico. So, even though we would be outsourcing the operation in Mexico for this flight, as well as future ones, I needed to ensure for this flight that we had all the kinds of skill sets we would need if we encountered any problems on the way.

Like Noah's Ark, I arranged to have two of every skill set on our flight that might be needed to handle any problem.

Most personnel were supervisors from their departments, so they were highly experienced and skilled. I had two sets of pilots, each with plenty of available flight time. I had an extra crew of inflight attendants. I had two airport agents in case questions about that were asked. I had two food service handlers, two aircraft mechanics, and two radio mechanics. In addition, we carried a variety of spare parts.

Before departing, I arranged hotel rooms for all those traveling. In addition, I prepared and loaded enough airline paper materials for the first six months of operation.

When the day arrived, and the aircraft was loaded with passengers and ready to depart, we encountered our first problem. The food service personnel who delivered the food buffets to the aircraft in coach class needed help pushing them into their places. They usually slide into an opening along tracks, and one of the tracks had come loose, and the buffet jammed. We had not yet taken a delay from our planned departure time, but we were at risk of doing so. So, I asked the two food service supervisors we had on board to get involved, which they did. They quickly got past the problem and locked the buffets into place. Whew. Problem solved. The plan was still on track.

Next, during the flight, I got word from the cockpit that the pilots found while flying over the Gulf of Mexico that the frequency they usually use for communicating with air traffic control changed when they got out of range from the U.S. and into the range of the Yucatan. They figured out the right frequency before getting lost or causing consternation with Mexico's traffic control people, and the FAA inspectors and passengers were none the wiser. However, in the process, they uncovered another problem with their radio.

Landing in Merida, we were greeted by the mayor and a mariachi band. Everyone got to stretch their legs and be entertained with music and speeches from their mayor and our president for a total of about a half hour.

Moving on to landing in Cancun, I was shocked as we taxied to the terminal building to see about forty armed soldiers pouring out from behind the terminal building and forming a circle around the aircraft. Momentarily, I worried that someone had reported my illegal shipment of collateral. I was the first to get off the plane so I could greet our ground operator, who was waiting for me. He then yelled a few commands in Spanish, and all the soldiers did a sudden "about face," looking away from the airplane. Once they were not looking, his ground crew appeared and started to unload all the contraband collateral material I had "imported" illegally. He took it all off to some house he owned nearby to store it until needed.

While the unloading was progressing, all the passengers exited the airplane and went to the hotel by shuttle to relax and dress for dinner. I remained behind in the cockpit, watching our pair of radio mechanics try to fix the aircraft's radio system, which was still on the fritz. If they did not get it working, we would have to order another plane to fly down to pick us up, which would not sit well with the FAA.

I watched the radio experts try an endless combination of settings for a set of switches. They explained what they were doing as they tried each combination, but nothing seemed to work. Finally, I noticed one combination that they had not tried and suggested they try it. They mocked the idea initially, saying it was an illogical combination. Still, after trying the settings that were supposed to work again without success, they finally agreed to try my combination. The combination worked. The wiring had been crossed when the switches were installed. Whew! Another problem was averted.

After arriving at the hotel, I checked in but did not have time to change my clothes, even though I was already appropriately dressed. I waited for the guests to arrive in the reception area, where we planned to start with cocktails before dinner. The first two couples arrived shortly after.

Our ground operator had invited all the local hotel owners from Cancun and Cozumel to the party, so we expected around 150 people. Cozumel is an island 50 miles southeast of Cancun, accessible by ferry or airplane. Because of the distance, the hotel owners from Cozumel had arranged hotel rooms in this hotel, anticipating they would not return home that evening.

While waiting, I struck up a conversation with our company photographer who was one of the PR people whom I had invited, when I overheard two couples arguing loudly nearby. They had been drinking, and were getting rowdy. Curious, I joined in and asked what the dispute was about. It turned out that both couples were hotel owners who knew each other and the owner of the hotel we were staying at quite well. They had both requested a suite for themselves, but only one was available, which meant that one couple would have to settle for a standard room, which would be below their expectations. Since they could not come to a resolution, I suggested they flip a coin to determine who would get the suite. To my surprise, neither couple had heard of this method of settling an argument, and they found it amusing. They eventually asked me to explain how the process worked.

I pulled a quarter out of my pocket and showed them the heads and tails sides, explaining that one of the ladies - pointing to one of them - would choose which side they wanted. Then, I would toss the coin, flip my hand over, and place the coin on my wrist. If the chosen side was facing up, the lady choosing would get the suite. They both agreed that the process was fair and practical, so I flipped the coin. When I revealed the coin, the victorious lady jumped at me and kissed me on the cheek, leaving a bright red lipstick mark, which the photographer was waiting for. I was a little worried that the image might appear in the company's internal newsletter, but thankfully it did not. However, I did receive a print of the photo from the photographer in the mail a week later. The rest of the party went smoothly, and everyone had a good time.

The following morning, the United team flew to Cozumel to attend a scheduled breakfast. The purpose was for our management to gain firsthand knowledge and understanding of these destinations, so they could speak about them with authority. As we walked into the restaurant for breakfast, a couple of executives, anxious to return to the office, accused me of going too far with the planning and wasting their time. They believed I had crossed the line from essential work to unnecessary extravagance. I pleaded with them to be patient.

The weather was beautiful, with balmy breezes, and the company of local businesspeople was cordial. We sat overlooking sparkling green water teeming with marine life. The food was superb as well. After breakfast, the same executives came to me again and said they were mistaken, that the experience was worth the time, and that the whole trip was well executed. I had so much help from people all over the airline that it was not for me to take credit alone, so I shared their comments with all those who helped me after our return.

Later, I would serve a lesser role in starting another new flight service from Chicago to Tokyo, Japan. I flew to Tokyo on JAL and flew back on Singapore Airlines, with the objective of seeing what their first-class food services were like. In Tokyo, I also had an assignment to meet with Victor Jules Bergeron, Jr., well known as Trader Vic, a famous restaurateur about possibly providing the inflight meals for the flights from Tokyo to Chicago. We met at his restaurant. The food was good but not remarkable and he struck me as eccentric. I do not recall who got the catering contract, but it was not him and my input had no bearing on the decision as I recall.

Boeing 757

Another project that came my way was to visit the Boeing aircraft plant in Seattle and look over the new 757 that they were developing for any operating performance improvement opportunities.

I was excited to witness the creation of a new model of airplane. In their process, they first built a prototype from plywood. Then they use that full-scale model to refine the engineering drawings for building another full-scale model from metal and composites. Lastly, they make the first production model. Then, oddly enough, they string that airplane up in the air by its wings inside a unique hanger and bounce it up and down until it breaks. That way, they find the most obvious structural weaknesses.

I also learned another unique fact about building airplanes. What do you think would be the first part brought up to the production floor to build a plane? You probably would guess the wings, but that is not it. The first part is called a "wing box," like the bottom of a gift box: four sides and a bottom, but no top. This part is welded from thick, sturdy metal plates. The wings are then laid on the box and bolted firmly to that part. Then the plane's body goes on top of the wings and the wing box.

When I visited again, after they had built the first production model, I found two important changes that would affect the plane's operating performance. The first was simple to point out but not easy to change. The galley trash container was stuck in a corner and could only be removed for emptying by moving the food service buffets. This would not be possible on a "through" flight, where schedules allow 20 minutes. They agreed and later redesigned around that problem.

The other even more important change was that I had calculated the travel time of the airplane from one coast to the other and back again. The fueling time of this critical round trip was too long to allow the plane to fly out and back on the same day. This really would limit the use of the airplane. As a result of my observation, they added the ability to put fuel in from both wings rather than just the one they had planned. This crucial modification allowed the plane to be used as intended and not become an albatross in the industry.

End of Term as Chairman

While in this job, I applied Jim Kent's Management Information System to document all job aspects, such as policies, guidelines, procedures, standards, and checklists.

The resulting binders filled the entire six-foot-long credenza behind my desk chair. When I was promoted, on the morning when my replacement arrived, I instructed him to read every binder from left to right, and I would answer any questions upon my return from lunch. Upon my return, my replacement informed me that he had finished reading all the binders and had no questions. He also added that it was the most comprehensive job turnover he had ever encountered. All thanks to Jim Kent.

At the end of my three-year tenure as Chairman of the Operating Performance Team, we received an award from the president for achieving the highest three years of operating performance in the company's history. I say "we" because without the support of my boss, Herb Hubbard, and all the committee members, this success would not have been achieved. Several team members displayed political courage by taking stances that were beneficial for the company, even when they went against explicit instructions from their bosses or division heads. I witnessed their support through votes, as they stood up for the team's actions because it was the right thing to do.

Reservation Offices and Ticket Offices

After three years in the high-pressure Chairmen position, I became the Manager of Reservations and Ticket Offices. The title of "Manager of" meant I would be managing managers. This new job called for me to manage the processes and improve the efficiency of the 2,000 employees across the company's twelve Reservations and ninety Ticket Offices. This responsibility included operating procedures, policies, ticket formatting, exchanging travel privileges with other airlines, performance goals and productivity, and much more.

The expectation was that I would lead the consolidation of Reservations offices into a smaller number of larger offices resulting in a reduction in force. I immediately began to question my assignment.

I received an abbreviated turnover from Bob Gormley, my predecessor in the role. The only advice he gave me was: "When you open a retail location, such as a ticket office, only select the location based on the amount of foot traffic that passes in front of it." This was and still is a basic yet vital lesson for locating any retail business. This concept was the basis for the original emergence of shopping malls. More traffic walking past means more potential customers.

I inherited a few very experienced managers and hired two new young, energetic, inexperienced managers. They all sat at desks in a bay in front of my office. I could see the new hires through my doorway, but the experienced guys were out of my view to my left. That was okay because they were hard workers, knew more about what they were doing than I did, and did not need me to oversee them.

The Reservation Offices accommodated a significant number of agents, ranging from 65 to 200 per office, which required ample office space. The workload, including phone calls, schedule changes, interactions with travel agents, and ticket mailing, determined the required number of agents. Consolidating the offices would not reduce the workload or the number of workstations. It would result in the loss of experienced agents who could not relocate and their replacement with inexperienced ones. Moreover, I failed to see how physical consolidation would lead to significant management expense savings, as the supervisor-to-agent ratio would remain constant. Thus, the idea of consolidating offices appeared illogical and lacked the typical conditions for achieving Economies of Scale gains.

So, I decided to electronically consolidate the workload rather than the people. It was the balancing of workload across offices across different time zones where the efficiencies were to be made. If one office began getting more calls than it could handle, we would "overflow" those calls to an office where agents were idle. That way, the agents would be more fully utilized, and the company would benefit by eventually needing to hire fewer agents.

The systems that managed inbound calls from travelers were called Automatic Call Distributors (ACDs for short). Rockwell Collins made the ACD system used by each office, and it was electro-mechanical and difficult to work with. The current state of ACD technology was primitive, and with the incursion of computers into the ACD business, companies were beginning to build computer-driven systems to handle inbound calls.

I found a company in California called Aspect Telecommunications, founded by Jim Carreker, that was beginning to sell a computer based state-of-the-art ACD. I collaborated with Jim to enhance their ACD system's capabilities, enabling callers to leave their phone numbers when all agents were busy, and still retain their place in line, receiving a callback as soon as an agent was available. Although this functionality is common today, it was not available until we developed it. I created a flowchart of the process, and Jim integrated it into the system.[3]

During the early 1970s, AT&T charged by the minute for inbound call telephone lines. Every minute a caller waited on the telephone line to talk to a reservation agent cost the airline money. So, capturing the phone number and then calling customers back saved money and helped pay for replacing the ACDs. We also interconnected the ACDs between all the offices so that if an office got too many calls, they would overflow to Reservations Offices where agents were available. Distributing calls across all the offices worked well and increased productivity, requiring fewer agents over time and keeping all the agents busy more often. No one got laid off, and the original goal to save money by consolidating was achieved.

I had one manager, Bob Camastro, who was very bright and had great potential for advancement. I wanted to challenge him so he would grow in his job quickly. So, I decided to assign him a project that I knew he would not have the experience to handle -- would not have any idea how to do it. After I dropped the folder on his desk, I walked back into my office and watched him through my doorway as he opened the folder, studied its contents, and then sat there looking bewildered. I let him stew for a while.

I then walked back out and suggested he talk with the more experienced manager who sat right in front of him and ask him for advice on getting started. Visibly relieved, he did so and was soon off and running. He went on to be very successful in business, became CEO at several companies, and we remained close over the years, seeing and talking to each other as recently as a few years ago.

One of Bob's successes was a project to change the Reservations Operation from a service function to a sales function. Senior management was skeptical and denied our broad request but agreed to allow us to test the concept with the San Francisco office. Our plans called for offering a financial incentive to the Reservations Agents for selling hard-to-sell flight options. This would mean reorienting some agents from being "order-takers" to being active salespersons.

When we first proposed the idea of agents selling, an uproar ensued, with complaints across the offices that "we were not hired to be salespeople. We were hired and trained to provide service." We agreed and let them know their participation was optional and they could stop anytime. The proposition was to book customers on United's one-stop connecting flight between San Francisco and Dallas, TX. This option, despite being slightly more expensive and arriving later, would be offered rather than other airlines' non-stop flights to the same destination. We offered the selling agent $5.00 per booked seat.

In a week, a small portion of the 200 agents sold 1,300 seats on flights that would have otherwise gone empty, resulting in nearly $300,000 in additional revenue for the airline. The total cost of the incentives paid to these agents was $6,500. If all the agents in the office had participated (which was not the case), and each had made an average of 6.5 bookings, they could have earned an additional $32.50 per week. Over a year, this would have amounted to an increase in pay of $1,690 per agent, or about 10% more than their current earnings. Not all the agents participated so those that did earned much more than this. If the program continued, it could potentially have resulted in an additional $13 million in annual revenue for the airline. And this was based on a single flight opportunity.

The test was a roaring success no matter how one looked at it, except with senior management. Senior management decided against continuing or expanding the program without providing specific reasons. However, I presume their concerns included potential customer perceptions of being taken advantage of and the potential negative impact on agents who were unable to transition from service to sales, as it could create a division between two classes of employees – service and sales. Also, the program would take active management, and the agents who succeeded would make more annual pay than other comparable jobs in the company. Finally, the program was just too radical for executives who preferred to avoid taking risks.

After I served a couple of years in this position, the west coast president was promoted to president of the airline. It was a bad omen for me as he disliked me, mainly because he knew I had no respect for his management style. While he was a VP, I once witnessed him inform a manager that his next paycheck would be delivered in Pittsburgh and that he needed to be there to receive it. This was his way of unilaterally transferring a loyal and hard-working employee who had a family and a home without any company-provided relocation assistance. The VP could sense my disapproval from my body language.

I had a few other analogous run-ins with him, so I knew my days were numbered when he moved into the president role. He became president in 1981 and CEO in 1985. He served as CEO until 1987, and he remained chairman of the board until 1988.

Ticketing Devices

During those days, tickets were transferable between airlines. Tickets were printed by mechanical ticket printers, like dot-matrix printers. The long-term vision was to have ticket readers at the check-in counters in airports. For these reasons and more, the format of tickets had to be highly exact and standardized across the airlines. The one person in the industry who defined those standards worked for me. It was a full-time job to develop and maintain the standards because different airlines had different needs, and not only did he need to accommodate them, but he also had to negotiate them. The airline also had a quiet guy in the computer center's basement in Denver who built ticket printers by hand. He was an amazing unsung hero. Each time his design revision worked for the latest standards and were connectable to Apollo, that design would be sent to a manufacturer to produce them in volumes.

Maintaining relationships with the other airlines fell into my purview too. For example, I would issue flight passes, free or at a discount, to executives at the other airlines. Likewise, I was the conduit for requesting passes from other airlines for our executives and other "deserving" management personnel. No standards or policies existed, so I had a lot of discretion. I did not try to create any either because there were so many variations and nuances.

This was a time before email and the Internet. Our means of communication across the airline was called "meters." Meters were like telegrams and were addressed with a five-letter code. Ours was EXORZ for Executive Offices Reservations. Different departments in different locations had their department codes, like LAXFO for Los Angeles Airport

Flight Operations, and so on. We also had group codes that made it easy to send new procedures related to reservations and ticketing to all offices in the airline. Our messages were always signed EXORZ-Patton. That address was widely known across the airline, almost infamous, but certainly widely recognized. Hardly anyone across the airline knew my face, but they all knew my department and last name.

Emergency Task Force

One of my responsibilities was managing a portion of the Emergency Task Force, which operated independently from the formal company organization and handled all aspects of an aircraft accident or emergency. It was like a secret society, like the Knights Templar. If an accident occurred, many tasks had to be completed, such as dispatching recovery personnel to the accident site, establishing family liaison centers, providing counseling, coordinating with FAA investigations, and even transporting body bags to the scene. The airline, rather than the local government, managed the crash site. Over time, specific individuals were assigned various roles within the airline's emergency response team, regardless of job title or rank.

My assignment was to operate a team of six associates whose job was to determine which passengers were on the flight. This sounds simple, but it is not. In those days, passengers might have boarded at the last minute, connecting passengers might have missed their connection, there might be some "lap children" on the flight, and so on. The main source of information would be the stack of tickets collected at the airport(s). But we also had the reservations records. Also, there were the pilots and inflight crew members, some of whom might be "deadheading," flying back to their domicile city.

My team's job was to go to an interior room in the basement of the executive office, remove from cabinets computer terminals and printers and get them operational, then contact the airports in the line of flight, pull all passenger manifests, and build a consolidated passenger list that was 100% accurate. Mistaking one passenger could be devastating to that person's family.

As soon as the passenger list was ready, I was to rush it to a command center where the president and many of his senior vice presidents would be sitting around a table on phones, talking to the FAA, the press, and others. Because the process was intricate, sensitive, and urgent, upon an incident happening, it was my job to make sure it all went smoothly, 100% perfect, fast, with no hiccups. As a result, I drilled my team on the entire process religiously every month, despite repeated complaints that they did not need to drill any more. I was hoping I would never see it in actual operation, but I was diligent.

Well, one day, I got "the call." An aircraft had been hijacked and was reportedly headed to Cuba, or so we were informed, and I was told to activate our center. My team was so well drilled that everything that needed to be done went like clockwork. Our standard was to have a passenger list ready within 20 minutes. I had it in my hand in 15 minutes and rushed to Command Center.

When I arrived, the room was like a beehive of commotion, with the president "large and in charge." I walked up behind the president, looking for an opening to hand him the list. At the instant I arrived behind him, he bellowed to no one in particular, "Where is Patton with the damned passenger list?" Someone in the room said, "He is standing right behind you." He snatched it from me, asking if there were any "notables" on the flight, meaning dignitaries, government officials, etc. I told him, "Not that I recognized." As it turned out, the flight returned to Miami with the "hijacking" averted. The incident, to my knowledge, was never reported either internally or externally. I was relieved and pleased that my team performed flawlessly.

Overnight Package Delivery

I recall one almost humorous situation that averted a major embarrassment for one guy. I stopped in to visit the latest VP of Cargo. The SVP of Marketing had a practice of rotating all his direct reports every December. His goal was to broaden their experience and provide some "cross-fertilization" of ideas. So, I asked the Cargo VP what was new in his realm.

He told me they were about to start a new service in two weeks to compete against FEDEX with overnight package delivery. I was shocked when I heard that because I had not heard a word about it. Thinking quickly, I asked him, "And who will handle your phone calls?" He looked at me quizzically and said, "What phone calls?" I realized then that he had not thought about the need to take calls. I reminded him that people would need to call to order pickups and inquire about their deliveries. He gasped. He had missed this significant component of his new service. However, he did tell me that he had the systems ready to take orders. I told him my team would scramble to provide what he needed.

I knew that the Detroit Reservations Office had a vacant second floor. We shipped cubicles from other offices, got their ACD set up to receive his phone calls, got computer terminals shipped and installed, and assigned a local supervisor who rushed hiring and transfers to set up the office in only a few weeks. We accomplished what the VP needed in time for his major announcement and start-up of the Package Delivery service.

Unfortunately, the service would close after a year of trying to gain enough market share to make it profitable. This showed the power of my team in solving such a complex problem against a crazy timeline. My team knew how to "work the organization" to get things done. Not everyone in a big organization is willing to take quick action and get things done. It takes people who have the trust of their bosses, informal connections across formal lines, thick skin, and willingness to take "career risks" to do what is right for the business. People will take risks but only if they know their boss has their backs and will take the hit if the risk fails or if they operate with their own set of uncompromising principles.

Mileage Plus

On May 1, 1981, when American Airlines announced its loyalty program, AAdvantage, United Airlines' management went into a tizzy. They hated following any other airlines' lead, but America was often the industry leader. From a historical perspective, I recall that when United wanted to add an

Accounting System to its travel agency automation offerings, they leased a system from a company in Atlanta. The next thing that happened was that American Airlines bought the company from which United had leased the system. System support, as I recall, slowed measurably. This and the fact that Apollo was based on an Eastern Airlines copy of American Airlines' Saber system further reflected that United was a follower rather than an industry leader. This annoyed management.

So, when American announced its AAdvantage program, our Marketing Department, of which my department was a part, was ordered to do all that was necessary to match American's offering as quickly as possible. A team was assembled in a conference room and instructed that they were not to leave the room until they had designed a product to compete with American's AAdvantage.

Our first challenge was to evaluate the underlying economics of a rewards program. It was evident that such a program's primary advantage and threat would be to lock customers into flying repeatedly with the same airline, which would be the one with the most lucrative rewards. Any airline not offering rewards would lose customers. It became evident to me and others that if we did not align our program with American, United would likely lose customers to them. Additionally, United could potentially attract customers away from American Airlines and other airlines, thereby offsetting or even surpassing the initial loss. Considering that United had more flights than all carriers except American, the decision was clear. We had to match them or do even better.

The second challenge was determining how much "reward" would make economic sense without "giving away the store."

We used American's model as a guide and calculated the cost of their program, allowing us to match it.

Thirdly, the program needed a name. Although I do not remember who specifically came up with the name MileagePlus, it was quickly approved. After outlining the program and gaining approval to move forward, we spent the weekend developing detailed plans for tracking United's mile rewards, determining the redemption process, designing necessary collateral materials, and establishing how customer service would be supported. Eventually, a separate reservations department would be needed. The program was announced soon after and implemented within a few months. Both AAdvantage and MileagePlus are still active loyalty programs today. I think if airlines had their druthers and could trust all airlines to follow suit, they would close their rewards programs. But that will never happen.

The End Cometh

After my two recent successes, delivering the passenger list on time to the president and saving the VP of Cargo's career, I hoped my position with the president might improve. I had been in my current position for three years and was ready for a new challenge. During a conversation with Howard Putnam, the SVP of Marketing, I expressed my interest in an open director position in his division. Although he was willing to promote me, he candidly admitted that he knew the president would not approve it. Despite this obstacle, I appreciated Mr. Putnam's honesty and recognition that I understood the reality of the situation.

I had not entertained leaving United up until this point in time, but it soon became clear that my involuntary career change was imminent when the CEO soon moved one of his old west coast cronies into a position over me. And this VP, Ernie L., was candid enough to tell me directly that he had been instructed to replace me and transfer me into a job with no duties. By not firing me but only boring me, the CEO figured he could avoid an age discrimination lawsuit since I was now over age 40. The CEO knew me well enough to know I would not stand for doing nothing and figured I would quit.

To demonstrate our team's exceptional abilities, I rallied my managers to give their utmost effort and perform at their best for the next week or two. I encouraged them to go above and beyond in all tasks assigned to us, with the aim of impressing Earnie L. and showcasing that we were the most competent and skilled team in our respective roles.

We did so, and Ernie L. was dragging his feet on making the change. It was as if he were allowing us time to show what we could do. I wanted him to know what a mistake the CEO was making.

When he finally called me into his office to tell me my fate, he admitted that he was impressed with our performance. But he told me he had no choice but to move me out. I knew that my time was up at United Airlines.

What I learned at United Airlines

I learned various programming languages, Fortran V, ACP, and FOCUS. I learned how to design, develop, and implement large, nationwide computer systems using dedicated data lines and how large computer centers are built and operated. I learned how to deal with Big Data. I developed the first diagram of the Computer System Development Process, which would inform my future project developments.

I learned about ACDs and more about data communications lines. I learned how to manage large groups of people, like 2,000 reservation agents, operating from a staff position and working through multiple levels of management. I also learned how to electronically consolidate call centers.

From Eddie Carlson, Jim Kent, Percy Wood, and Barbara Allen, I learned what good leaders look like and how they lead. I learned from a few other executives what bad leaders look like. Finally, I learned how to get big projects done within large political organizations.

I personally set a goal to observe and learn from senior business leaders, studying their management styles, their behavior under pressure, their treatment of people, and their level of expertise in understanding the intricate details and interrelationships of the business. Knowing how Senior Management functioned would prove valuable to me later in proposing and negotiating to get funding for major system developments.

I learned about industry regulation. I was once sent to Switzerland to represent United in an official IATA proceeding, but with the strict instruction to listen and say nothing. I also participated in several FAA Proposed Rule Making responses.

I learned to seek input and trust in the experience of people working at all levels of an organization.

While I was there, I developed a parable for use when counseling those who wanted to know how to get promoted. It paints a picture of how promotions work. Imagine yourself standing on one side of a wide crevasse, looking over to the person who can throw you a rope and help you cross to the other side. But here's the catch: if that person does not like you, dislikes how you dress or wear your hair or your piercings or believes you will not fit in with their circle, they might choose not to throw that vital rope. As part of this process, you should consider how they think you behave and how they think you should dress, not how you think you should dress. They've got to like you and see that you fit in with them.

I was fortunate to have held three jobs at United that gave me free rein to learn how the entire airline functioned. I learned the intricacies of an airline, including how to build an airplane, how airplanes are maintained, how schedules are made, how food service functions, how Reservations Offices operate, and much more. I experienced the widest perspective on the whole airline, more than anyone I knew in the company. It was their loss that they failed to recognize my potential for a higher office. I could have run an airline and, in fact, did on a small scale, as you will read further on.

Firsts at United Airlines

Several firsts were achieved at United during my tenure:

1. The creation and implementation of the global airline reservations system, Apollo, which is still in use under the name of Galileo.
2. The creation of the team that created and implemented the Food Service Division's comprehensive and fully integrated Food Accounting and Management Information System (FAMIS).
3. The first nationwide operating performance daily tracking and reporting system and the establishment of goals for every operating department across the airline.
4. The achievement of the highest three years of Operating Performance in the airline's history, with the help of the entire Operating Performance Committee. Including:
   a. Recognizing the importance of STAR flights and implementing nationwide goals for their performance.
   b. Recognizing the importance of scheduling on operating performance, producing the first operating performance report by aircraft routing, and providing guidance to the scheduling department that permanently eliminated many built-in delays.
5. The first fully automated president's Operating Performance Pocket Piece.
6. Member of the team that designed, over a weekend, the second Airline Point-based Rewards System – MileagePlus, and then implemented it in less than three months.

7. Establishing the customer service operation for the start-up of overnight delivery service.

8. The creation, with Aspect Telecommunications, of the first system capability to capture a caller's phone number, keep them in the queue in order, and then call them back as agents freed up.

9. The first (and last) test of incentive-based sales in an airline reservations operation.

CHAPTER 3

ENTREPRENEURIAL ENDEAVORS

Crescent Helicopters

After leaving United, I hung out my shingle as a consultant. I did have an MBA from a prominent business school, so why not become my own boss? I would work in four fields: Airlines, Oil Refining, Call Center Reservations, and Sports Medicine. Unfortunately, my immediate expertise applied well to only one of these. The others would be learning experiences. I started with airlines.

Soon after leaving United, I was introduced by a friend to J.P. Sauerman, the CEO of Sauerman Inc. The company, called Sauerman Inc., manufactured large, earthmoving scoops known as "Sauermans." It had been invented by his father and improved over many years to the point where it was the top-of-the-line dirt scoop for a dragline.[11] It was used for large earth moving projects like building dams.

At one time, the company had a large department of engineers. However, because their main product was so optimal and durable, resales eventually replaced new sales, and the design of new products died out. So, when I met J.P., his manufacturing area was manned only by a few machinists, who repaired buckets, made new teeth, and sharpened old teeth. During the year I was there, they sold one new bucket for $1 million with a 50% gross margin. They also owned three helicopters, originally for visiting worksites, but at this time, they used them for charter flights, such as flying transplant organs between hospitals.

J.P. wanted me to convert his fleet of helicopters into a Chicagoland Airline called Crescent Helicopters Inc. that would fly between Meigs Field in downtown Chicago, Gary Airport in Gary, Indiana, where the company had a hanger, and O'Hare airport. They would also fly charters to the airport in St. Charles, IL. He wanted me to pull the helicopter operation into an airline as its president. To say this was an ambitious goal would be an understatement. With much work, I believed I could do as he wished, gain great experience, and the title would look good on my resume, so I agreed to a one-year contract and abandoned other consulting opportunities for the interim.

He had two pilots. The first and lead pilot was an ex-bush-pilot for an African Missionary operation. The other was a young man with limited experience, flying for the U.S. Army in Vietnam. The lead pilot thought he should have been president and held some rancor toward me while I was there. He would not have known how to do what I did, but neither did I have the experience he did.

The company owned three helicopters: a Bell 500, a Bell 300, and a German Bölkow. The Bölkow, which cost over $1 million, was acquired on a lease payment plan. During this period, I built a model of the company's profitability as an airline. Based on my analysis, it became apparent that the expensive helicopter would never be able to pay for itself.

I informed J.P. that we needed to get rid of the expensive Bölkow helicopter. He did not think we could because it was on lease. He asked me how we could do that, to which I replied that I would call the leasing company and inform them that we would no longer be making payments, and they could take the helicopter back. I made the call, and to P.J.'s surprise, the company agreed to cancel the remaining balance on the lease and took back the helicopter within two days. This move relieved a significant drain on our cash flow. Unfortunately, the Lead Pilot was unhappy about losing his prized possession, but it was necessary to improve the company's finances, which were negative even without the helicopter.

At one point, J.P. offered to sell me the whole company, including the metal fabrication portion, for $1. I said no thank you because the company carried heavy debt on both the machinery and helicopter sides of his business, the machinists were all approaching retirement age, and the machining tools were all old, although still serviceable. If I bought the company, I would spend the next year liquidating it and maybe walk away with $100,000. So, I continued to try to create a profitable business for him,

I registered the company as an official airline with International Air Transport Association (IATA). This was essential if our schedules were to be loaded into the Travel Agency Reservations Systems. I knew how to get our fares and schedules into the travel agency systems when ready, but we needed an airline code, which I obtained. I do not recall the code, and I could find no official record of it.

Keeping the operation running was a negative cashflow nightmare. I would call our accounts receivable a few days before every payday to collect enough money to make the payroll.

I had to arrange a place to land at each airport before I could finish the schedules. Meigs Field permission was quickly achieved because we already flew in on charters. Landing in Gary, IN, was fine. They welcomed the business. I then negotiated with the City of Chicago to operate in and out of O'Hare airport. I wanted permission to land atop the parking garage in front of the terminal. That would have been ideal. Next, I began working with the Small Business Administration for a loan of $500,000 ($1,300,000 in 2023 dollars). I had everything lined up at one point, but SBA would not budge until the city approved our plan. I could not get the city to bend, and they eventually said no, sinking the loan. So, I arranged to sublease a shared-terminal gate at O'Hare. It was distant from the major airlines' gates but still within walking distance.

I developed a fare structure and a schedule that would optimize pilot availability. When we finally launched, only a few bookings happened. The main reasons for our difficulties were a need for more awareness among Travel Agents, limited demand for helicopter transportation, and the high fares required to cover our costs. I found we needed to continue marketing our charter operations with hospitals (for organ transport). That kept us in business but still losing money.

I recall one hospital charter where the Lead Pilot asked me to approve them flying with the helicopter when it was out of FAA mechanical compliance. He had repaired a part that should have been replaced. I refused to let him fly the charter, and he was furious with me, but I stood my ground. I would not be responsible for someone's life having ignored FAA Regulations.

On another charter, the younger pilot flying a passenger from O'Hare to St. Charles ran out of fuel and had to land in someone's backyard in St. Charles. The passenger thanked him for landing close to his house and walked home from there. But, again, I was not happy with the pilot or his boss.

On my first anniversary, I looked over the books and realized that the business would finally break even if I furloughed myself. I felt it could run itself at that point, so I departed. However, it had poor prospects for further growth and would eventually fold. No record of Crescent Helicopters remains in the IATA database.

Petroleum Processing

During my time at United, it was common practice for managers to fly first class on business trips. On a business trip to New York, I found myself in the roomy first-class cabin on a DC10. However, I misread my boarding pass and accidentally sat in the wrong seat next to the only other passenger in the cabin. As soon as I realized my mistake and stood up to move, he kindly told me to stay put since I was already settled.

I introduced myself to him. His name was George Schlowski. I noticed he was reading a book on computer programming. I asked him what he did for a living, and he told me he was a consultant in the oil processing field. I told him, "George, you are undoubtedly smart enough to learn how to program, but why not hire someone who could do that for you much more quickly?" He asked if I was volunteering, and I said I would be willing to help him if I could, but not for free. He hired me on the spot.

This chance meeting resulted in a "beautiful friendship" lasting for years. I programmed several oil processing applications for him in my spare time. He worked with refineries, in Texas. Their processing was primarily turning contaminated oil into usable chemicals. My programming primarily involved performing mathematical calculations to determine the compositions and volumes of liquids in large containers of various sizes and shapes. These computations often relied on dipstick measurements of the contents' depths.

I recall one crowning achievement when he had given me a textbook containing the mathematical formulas for a volume calculation of a cylinder lying on its side with concave ends. As I programmed it, the formula in the textbook did not match the formula I had derived. I checked and double-checked my derivation, trying to prove myself wrong and the book right, but it was not working. Finally, I took the problem to George, who confirmed that the textbook was wrong, and I was right.

I worked off and on for him for a couple of years, including after I left United Airlines, and his business helped keep my family afloat while I looked for a more permanent job or built up my consulting enough to work for myself.

Reservations Staffing Model

My next job was to consult with small airlines and, in one case, develop a staffing program for an airline reservations office operation. While at United, we worked on a reservation office operation staffing model. After I left, I recreated, improved, and generalized it for universal use.

I convinced the president of a small airline in the Northwest to try it. As it turned out, he wanted me to do a broader consulting project for them, to which I agreed. I was invited to streamline the entire operation of their reservation office.

I worked on the project on-site with the local reservations' office manager and from home on my own for a couple of months. When I flew up to present the findings, the president called his entire management team into the room for my presentation. I had sent him my results in a report beforehand, so there would be no surprises, right?

I began presenting my findings, giving fair credit to the reservations manager who had worked with me. Before I had gotten more than halfway through my presentation, the president launched into the manager I had worked with, ripping him up one side and down the other for allowing the inefficiencies my work was pointing out to have happened. I was embarrassed for the manager. The president's behavior made me very angry because that was not what I had signed up for or my management style -- to attack a subordinate publicly in front of others. I never had another opportunity to work for them again but would have declined if offered.

At one point, while consulting, I traveled to Phoenix, AZ, on business. My wife, who was still working for United Airlines, was also attending a conference in Phoenix. She was involved in promoting United's Apollo Reservations System for use by Travel Agents. So, I arranged to take her to dinner and went to the conference site to collect her.

United had a backroom area for its people to take breaks. I was to meet her there. When I walked in, I saw a large photo of a group of people with a question banner across the bottom saying, "Do you know who these people are?" To my surprise, or I should say shock, I was one of those in the photo. It was a photo of Barbara Allen's group – the team that created Apollo. My first reaction was, "I know who those people are."

Then my second reaction was, "Oh my God, I'm now part of history."

SportsCare

After the helicopter airline and doing some consulting, I had the urge to get into my own business other than consulting. Unfortunately, as I described earlier, I found consulting to be a feast-and-famine business. When you are working on a project, you have no time to prospect for your next client, and after a project, you do not have the next gig lined up yet. So, I began looking for a business to acquire. I investigated acquiring an existing business but rejected it because it came with a partner I did not know and a single account representing 75% of their entire revenue stream (an automotive manufacturer). Lose that account, and the whole business would die.

I then responded to an advertisement I saw in the Wall Street Journal for a businessman looking for a partner in one of several businesses he owned. His name was Charles B. He and his wife flew into Chicago on a private plane, picked my wife and me up in a Mercedes rental car, and took us to a fancy restaurant for dinner. He intended to make an excellent first impression and he did. His deal was that we would put up $50,000, and he would give us half the business with him as a silent partner. I would run it.

The business, a start-up in Sports Medicine, would be in Scottsdale, AZ, a fine place to live with many residents living active lifestyles. He already had hired an operations manager, an educated and licensed physical trainer to handle the medical relationships and personnel relations with the trained staff. My wife and I decided to take the leap and try this approach to independence.

My wife and I decided she would remain in Chicago until I got the business up and operating. Charles B. immediately put $25,000 in the business's bank account, half of the money I had given him, to get started. To handle secondary details, I hired an administrative assistant. My other "already hired" associate had the title of vice president. In less than a month, my team of three had rented a small one-story office complex, furnished it, agreed on a business name, SportsCare, created collateral, lined up the liability insurance we would need, and printed business cards.

Next, we worked out the business's details, interviewing Orthopedic surgeons who would refer clients to us. The goal was to provide "recovery services" to serious and amateur athletes who could get walk-in treatment with a physician referral for physical therapy or with or without a referral for physical training. It was not legal to pay physicians for referrals, and they were not permitted to invest in the business.

The plan was for it to be a franchise operation with the first "Store" in Scottsdale. We began scouting for locations and quickly found the perfect place. We were to build out the "model" location in Scottsdale and work out all the processes and procedures there while the other two investors would start looking for sites.

The next two franchisees were already lined up to be in Minneapolis, MN, and Columbus, OH. During our start-up phase, we met with the two potential investors for those expansion locations. Meanwhile, Charles B. was starting up other businesses, like a carwash franchise, in partnerships with other investors.

Our first location would include whirlpool tubs, a small submersion pool for accurately measuring body fat, massage tables, etc. Everything was humming along. We identified a licensed physical therapist as the first store manager. We found a building for the first location and were ready to sign the lease for the property and its buildout. We had the equipment on order. I told my VP to wait to hire the manager until I got another capital injection, as our bank account was getting low.

I met a local "designer" named Jeffery Bruce, who I found could design anything. He was one of those brilliant artists who could work in any medium. He was working on sales brochures for an owner of hundreds of Arabian Thoroughbreds. He also designed geometric tent coverings for open areas. He offered to help me design the layout of our first store and helped lay out our little headquarters office. He was one of the founders of the Make-a-Wish foundation and talked me into helping them with their systems, which I did even though they were primitive.

We had been operating for the first six months on what amounted to half of my original investment. I went to see Charles B. to request more funds. He avoided me for a while, claiming he was too busy until I cornered him and started to press. At this time, not just from my pressure but from others who had invested with him in other ventures, his "house of cards" collapsed. He had squandered $millions. The rest of my money was gone, and 50 other investors were out similar amounts. He had been running a highly creative version of a Ponzi scheme.

I returned to our office in a muddle. I discovered that my vice president had offered the job to the potential manager, despite my instructions not to do so, resulting in that individual quitting his job. I felt bad for him until I learned he had been unhappy where he had been working anyway and would be able to find another job readily.

My wife and I had been house hunting. Thankfully, we had not decided on one. My VP and I spent the next month trying to raise capital independently to save what we had built because we believed it had the potential to succeed on its own merits. In the process, we got an excellent introduction to the world of venture capitalists and investment clubs (like the TV show, *Shark Tank*). We presented to many of them as far away as Sante Fe, NM. However, we lacked the "Proof of Concept," a working, profit-producing location. No one would invest in our business, even though we had no debt, a sound business plan, and everything in place to start up. Another drawback was that it was a time of massive interest rates – some offers would have been at 16% plus Prime (27% in total). Even though we believed we would start generating revenue within three to four weeks, we could find no rational investment deal.

A massive lawsuit followed, and numerous legal actions were taken against Chuck B. Our small part of his empire collapsed with him. I could not take the money remaining in the bank account, even though it had come from me, because I could be charged with theft. One smart thing I did while closing the business was to use the remaining cash to pay the IRS for the outstanding taxes. That way, they would not come after me personally, which they would have done. Our business was a great idea, well executed, and one of the few businesses Chuck B. had controlled with the potential to succeed.

It was discovered that Chuck B. had engaged in fraudulent behavior by selling the same business to multiple investors. A car wash operation, for instance, was sold to at least three different people. Although he had conned fifty individuals into buying various businesses, only my deal involved securities. In the end, his actions on SportsCare constituted securities fraud and resulted in his incarceration. After all the liquidations, the recovered funds all went to the lawyers. While it was a costly lesson, it taught me the importance of creating and managing a business carefully.

Back to Consulting

Once I was finished with SportsCare, Jeffrey Bruce introduced me to a guy who owned thousands of acres of land south of Scottsdale. He had made his fortune growing potatoes in Oregon. He had bought the land with the thought of growing more potatoes, but changed his vision to growing homes, as the Phoenix area was rapidly growing.

What he needed was a plan for the development of his vast holdings near Chandler, AZ, southeast of Phoenix. He was elderly but was able to fly me in his small airplane forty-five minutes south. I was nervous when I first got onto the plane with him until I noticed that while he was animated and talking to me, his hands were calmly checking over the various switches and controls. It was obvious that his muscle memory of how to fly was very strong. I did get nervous again briefly when he landed the plane on a dirt road, though he set us down gently.

What he wanted to know was when to begin developing the land. I was able to find annual government land maps for a 20-year period and measure the rate at which growth was moving in the direction of his land. I did notice that there was one crossroad of the only two dirt roads in his vicinity. He did not own the land around the crossroads. The closest town to Chandler, AZ, was Phoenix, fifteen miles away.

In the end, he was unhappy when my analysis showed that growth would not reach his properties for another 10 years. I did recommend that he buy the properties around that intersection. He paid me but was not happy doing so. I am confident I gave him solid advice. Today the area is mostly but not entirely built out with homes and golf courses 40 years later.

Thisco - Missed Opportunity?

Upon my return to Illinois, I was offered an opportunity by a recruiter to become the president of a start-up. The venture aimed to create a central switch to distribute hotel inventory, with sixteen major hotel brands each investing $100,000. The negotiations reached a critical point when I met with the key person behind the project, a member of the Pritzker family, a prominent Chicago-based wealthy family heavily involved in hospitality. Despite their eagerness to proceed, I was uncertain about the adequacy of the funding and uneasy about the fundamental concept, given the struggles faced by airlines with similar centralized systems. Additionally, I was concerned about managing the relationships to avoid any anti-trust issues arising. Also, even if successful, the potential for expansion was limited by having only sixteen hotel brands. So, I declined, which annoyed the recruiter and the person who made the offer.

The guy who did take the job, John F. Davis, III, described the first four months as "demanding shuttle diplomacy trying to corral the hotel companies into alignment."[10] I would have described it as having been hellacious. But, to his credit, he did get it done.

In an article on hospitalitynet.org, John described the purpose of Thisco, later known as Pegasus Solutions, as follows: "Murdoch's company needed an electronic way to book hotel rooms to support the creation of a new CD hotel catalog. They wanted to create the ability to electronically link the Global Distribution Systems (GDSs) with the hotels' Central Reservations Systems (CRSs) to sell CD-ROMs at the same time as hotels were looking for a single electronic interface to all seven GDSs." Despite the challenges, John eventually succeeded in this endeavor. Although I passed on this opportunity, I have no regrets, as I found my subsequent experiences far more compelling.

What I learned from Being a Consultant

As a self-employed consultant, I quickly learned two valuable lessons. Firstly, when working full-time on a project, finding the next job can be challenging. This often results in periods of cash flow constraints between projects, and it can be anxiety-inducing to wonder if you'll find another project after the one you are working on. I have often told young staff that when they are feeling overwhelmed with work, to do their best on what they are doing right now and then pick the best next thing to do.

Hiring a talented salesperson can keep you consistently working, but doing so comes at the cost of sharing the income. Thus, it's a tradeoff. Secondly, running a business requires a broad range of skills beyond your primary area of expertise. For instance, you must handle pricing, contracting, administration, billing and collections, marketing, advertising, promotion, and sales. These tasks also take time away from your paid work time. That is why consultants often charge two or four times the average hourly wage of a comparable permanent position.

Another lesson I learned was that with proper management, a small capital investment can sustain a business for longer than expected if spending is handled prudently. Care is essential to maintain control over the business's cash flow.

I learned a hard lesson about controlling a business. I should never have given over my money to Charles B. I should have insisted it go into a company account with at least my signature required to remove any funds from the account.

I also learned that becoming "your own boss" is illusory. Most entrepreneurs end up working for stockholders, venture capitalists, investors, banks, or someone else. There is always someone to whom you will answer.

As a result of the experiences above, I chose to look for a more traditional position.

CHAPTER 4

RESORT CONDOMINIUM INTERNATIONAL (RCI)

A colleague at United Airlines, whom I had followed through a couple of jobs, also had left the Airline. So, when I started looking for a new job, I contacted him for ideas. He recommended me to a recruiter for a position with a family-owned Indianapolis, IN, company called Resort Condominiums International (RCI). The company played a crucial role in the timeshare industry by enabling timeshare owners to exchange their vacation time with other timeshare owners across different resorts and seasons. This service addressed a common concern among customers who were reluctant to buy timeshare weeks at one resort and be limited to using only that resort year after year. RCI charged members an annual Membership fee and an exchange transaction fee. If you are not familiar with timeshare:

> If you are not familiar with the timeshare product, a timeshare is a vacation property ownership model in which multiple individuals or families own a share or a specific amount of time (such as a week or two) in a vacation property, typically a resort or condominium. Each owner has the right to use the property for their allotted time and is responsible for paying annual property maintenance fees. Timeshares are often marketed as a more affordable and convenient alternative to traditional vacation home ownership.

As the exchange business began to boom, Arthur Anderson advised the company's owners, Jon and Christel DeHaan, to hire a layer of experienced vice presidents under them. The couple had started their business using 3x5 cards in shoe boxes on tables arranged around their garage, but the volume of exchanges had quickly outgrown that approach. When I met them, they had upgraded to an IBM 4341 computer to run an automated version of their original manual system.

Their automated system was little more than the automation of their 3x5 cards. I was hired as their VP of Information Systems, equivalent to the more common title of Chief Information Officer (CIO), but with additional duties. I reported to Christel DeHaan, who ran the operational side of the business. Jon handled the Sales and Marketing side of the company. The other executives were the VP of Operations, Lane Howell, VP of Developer Sales, Larry Gildersleeve, VP of Marketing, Herb Alfree; and CFO, Bill Morris. Collectively, we composed the management committee that ran the company.

I soon learned that the company was losing money, $650,000, for the fiscal year that was ending and that we had a turnaround project on our hands. I also found that although the company sold services based on equitable exchanges, its core system lacked objectivity and consistent fairness.

When I arrived at the RCI offices in Indianapolis, IN, the IT managers I inherited informed me that the company's IBM 4341 computer had reached its maximum processing capacity and was at risk of becoming unresponsive to our telephone agents. This was my first week on the job, and I decided to independently verify their claims by reviewing a series of capacity reports provided by IBM. As it turned out, they were right, and an emergency upgrade was necessary. The same week, the company had planned a management outing at Marion Lake, south of Indianapolis, which I was expected to attend.

During the management outing at Marion Lake, the company's president gathered everyone around a large square formed by folding tables and asked us to introduce ourselves and describe our roles. When it was my turn, I introduced myself and then announced that I had a requisition for $120,000 that required a signature to upgrade the company's computer urgently. I made this announcement in an open meeting as I was unsure who needed to approve it - Jon, Christel, Bill Morris, or a combination. My comment received a bit of nervous laughter from the group, and I later learned that the management team often made such decisions casually. However, a $120,000 request was a significant amount for the company at its stage of development and for that group. The COO, Christel DeHaan, said she would address my request when the meeting ended.

As soon as the meeting was over, everyone started moving toward some houseboats on the lake. Christel was sitting at a table overlooking the activity, so I sat with her, pulled out my requisition, and asked her to sign it. She tried to beg off and proposed that we go for a swim with the others first. I told her I would be happy to go for a swim after she signed it so that I could call back to the office with the purchase order number. It was that urgent. She gave me a perturbed look but signed it. I made the call, joined a houseboat group, and went for a swim.

In the weeks that followed, it became apparent that the design of the existing system was obsolete, unreliable, and on the verge of inhibiting the company's rapid growth. In the interim, we decided to upgrade to a single IBM 360. However, having a single processor without any backup was risky, especially considering the large volumes of transactions the company was beginning to handle.

Furthermore, their processes around arranging exchanges were weak, and they had developed exchange rules around fairness that required human intervention, which were at times arbitrary and risky for the business in the long term. New application software was needed.

How Exchanges are Arranged

Timeshare Exchanges can be done in one of two ways, circular or free form. The first way amounts to lining up member desires until a complete circle is formed. For example, a timeshare member who deposits a July week in Texas wants an Orlando week in September, and a member with a July week in Orlando wants a January week in Aspen, and, for simplicity, a member with a January Aspen week wants a July week in Texas. By aligning all three requests, everyone gets what they want. This approach requires a nearly impossible degree of synchronicity and limits the ability to complete exchanges and charge fees.

RCI managed trade equity by classifying weeks as Red (high season), White (shoulder season), or Blue (Off-season). The Red could trade for a Red, White, or Blue week, White weeks could trade for White or Blue weeks, and Blue weeks could trade for Blue weeks. This basic rule could be relaxed if a week was available and close to expiring. The process was more complicated, but this gives you the idea.

CI took the free-form approach, allowing the deposits and requests to float freely in their system without requiring circular alignment. This approach maximized fee revenue but carried the risk that trades might not be fair. For example, they could not allow a member with a winter week in Maine to trade for a Christmas week in Orlando, to cite an extreme. At this point in the industry, all timeshare stays were transacted as full weeks. Later would come split weeks and then daily bookings based on Points.

In assessing the condition of their system, I realized that fairness needed to be automated using algorithms rather than relying on human judgment. I successfully persuaded them to let me redesign the system. The new system would assign points to the weeks deposited, using algorithms that considered different factors for the valuation. Factors included the demand for the location, the season in which the week falls, and the size of the unit, meaning the number of adult sleeping accommodations and total sleeping accommodations available. Other important characteristics were also considered in the algorithms. This approach would enable a fairer exchange through greater granularity.

The point value of the depositing member week then became like a bank account for that owner, who could then go shopping for the week they wanted, provided the point value of that week was less or equal to the balance of points in their bank. It was not an exact science, but it was much fairer than their old method and could be executed by the computer in real-time.

I managed to convince the owners to take this huge leap of faith and adopt an entirely new design. The timeshare industry was growing in leaps and bounds, and the company had to keep up. Of course, I took a huge risk with my career, but I had confidence in the design.

Building a New Exchange System

I began to increase staff and organize around this large development project. I had a few good managers, such as Chuck Priller, who was the current programming manager, Larry Darrah, who became my head of Computer Operations; and Dayne Dickerson, who was a key manager under Chuck Priller. I also had responsibility for a warehouse operation under their manager Gisela Reibel. She knew her business and needed no guidance from me. There are other names I should mention, but my memory fails me.

The company soon acquired a much larger main office and leased three floors of an adjacent office building to house their telephone agents.

I was allotted half of the second floor for my programming team, which would amount to a team of 120 at its peak. When I did a quick calculation, I realized that the number of cubicles that would be needed, given the current size standard, would not fit in the space. I also recognized that if I made the decision on how to 'shoehorn' 120 programmers into the allotted space, I could end up with a revolt on my hands. Therefore, I needed to involve my management team in the decision-making process. I showed them the allotted space, gave them a blueprint of the area, and proposed that they design the work area to fit their needs.

They did a great job and returned to me with a layout that accommodated 120 programmers. I then ordered the furniture. When it was set up, and I was looking over the new layout with my managers, I was not surprised when one of them said, "These cubicles were smaller than we envisioned." It was too late to make changes, and the sizes were their decision, so they agreed to live with their design layout. While I recognize that I had to be strategic in my approach, I did so to maintain morale by allowing my management team to make the decision.

As part of renovating the new office, I proposed building a large, secure computer center and installing three IBM 360s, one for development and running reports, one for the real-time operating system, and one as an instant redundant backup for the operating system. It was a multi-million-dollar investment for the company. However, during this period, a weird economic reality existed with buying IBM computers. For companies like ours, when you bought three new 360s at one time, IBM gave a considerable discount. So large that when the time came to upgrade to the next model of IBM 360s, we could sell our existing computers on the "used computer" market for more than we paid for them. IBM handled the used resales for us. In other words, we made money by keeping up with the latest models. Sheriff Berry in Orange County, FL, told me once that he used to do the same deal structure with Harley Davidson every year that they bought motorcycles.

The computer center we designed was a "work of art." It had raised floors, halon fire protection, redundant power, air conditioning, and 1,200 dedicated fiber-optic data lines. The room had a glass-paneled corridor down the center, so visitors could tour the room without being in the room.

The halon system is a fire-suppression system that, upon a fire being detected, fills the room with Halon, a gas that would smother any fire, especially electrical ones, but not kill people. The system was sold based on being safe for humans in the room, provided they left right after shutting down the computers, which took a few minutes.

When we tested the system, I recall asking the salesman who sold it to remain in the room for five minutes after it went off. I stayed in the glassed-in corridor so I could remain longer than five minutes and could watch for the safety of those inside and could keep time on how long it took to clear out the gas. It all worked fine, but I recall the nervousness of the salesman, who had not previously experienced what he sold. However, he was far more confident after his experience, so I did him a favor.

The Computer Room was built in the interior of the lowest level, which was partly underground to protect against the threat of tornadoes. We rerouted all the plumbing for the facilities on the floors above so that no liquids would ever leak through the room's ceiling.

We became AT&T's largest customer in the Midwest. and had them bring fiber optic lines directly from the master loop that encircled Indianapolis. The electrical was routed from two separate substations. We designed the operation to be as fail-safe as we could.

Or so I thought. While moving into the new Main Office, a landscaper, as we would later joke, "a guy with a baseball hat and a backhoe," dug up and broke all 1,200 dedicated data lines laid under the parking lot and into the building. That section of the lines was the only point of vulnerability for those lines. AT&T worked non-stop for 36 hours to splice all those fibers back together. With splices, they would never be as perfect as they had been, but we never noticed any dropped calls later either.

And that was not my only surprise. A few months later, while giving a tour to visitors, I looked out a downstairs window and noticed, for the first time, a large transformer. Sitting outside the computer room area. I called my manager of computer operations over and asked him what would happen if that transformer "blew" as they sometimes did. He was not sure and got on the phone with the Power Company. They assured us they had a replacement only a mile away and would only take an hour to replace it. The following week that transformer failed—so much for 100% uptime. As Robert Burns wrote, "The best-laid plans of mice and men often go awry."

At this same time, we began work on programming the new exchange system design. Our 120 programmers worked non-stop for 18 months to complete the system, including design, coding, testing, training development, training, and implementation. The Exchange system was written in COBOL, with some Assembler exits running under IBM's CICS TP monitor.[1]

One of the trickiest bits of logic in the system was the core algorithm that would determine the "value" of a deposited week that would be used as the internal "currency" for selecting another week of similar value. To be as objective as possible, we chose two consultants, each to propose an algorithm from which we would pick a "winner" based on how "subjectively fair" the matches were. We employed some of our most experienced users to judge the results. One source for the algorithm was Arthur Anderson (AA), and the other was my old boss in the Operations Research department at United, Herb Hubbard, whose analyses I respected. The two recommendations were quite different in approach, but both gave acceptable results so that we could have used either. We used the AA version because our president, Jon DeHaan, trusted that company.

I wanted to be certain, so I began looking for a way to simulate the exchange system using the new algorithm to determine the expected success rate for exchange matches.

MathWorks

One possibility was a new tool called MATLAB by MathWorks. Jack Little and Cleve Moler founded MathWorks in 1984 and, soon after, I became intrigued by the concept of software that could simulate real-time computer processes. MathWorks was and still is a software development company specializing in mathematical computing software. Their flagship product, MATLAB, is a programming language and interactive environment for numerical computation, data analysis, visualization, and algorithm development. However, MATLAB was not suitable for my specific purpose of simulating the likelihood of RCI Members getting matches. So, I programmed my own version using Fortran.

Reengineering the Corporation

I was invited to join an elite consortium of high-level IT professionals organized by an MIT professor named Michael Hammer, The consortium required a fee of $25,000 to participate as part of this elite, invitation-only group to learn about MIT's research in process reengineering and what other major companies were doing to manage processes.

Because I had developed a comprehensive and critical process for developing our major project and I felt it should work well on any similarly large computer projects, I saw potential value in further refining our design through engagement with the high-caliber people on the panel, all from companies larger than RCI.

We met over several weeks and discussed process reengineering in-depth. The organizer and his partner, James Champy, took copious notes. I do not recall them being actively involved in the discussions beyond listening to what we were doing and sharing a few tidbits about their research.

About a year later, a book was published that became a best-seller titled "Reengineering the Corporation," written by James Champy and Michael Hammer. Mike Hammer toured the country, promoting the topic and the book and getting high-paying consulting jobs. They NEVER acknowledged those who had provided them with much of their content, even though we had paid him for the "privilege" of us educating them. I believe they purloined my process reengineering ideas and concepts, but because I never shared my diagram or process manuals, I do not think they ever understood the full meaning of the process. I learned a lesson about intellectual property.

I also realized from this experience and others that choosing how you want to live your life is essential. Your choice of battles can significantly impact your quality of life. Guided by pride, I could have sued Champy and Hammer, accusing them of stealing my intellectual property. However, it's likely we had signed an agreement preventing such claims. Furthermore, proving copyright infringement is challenging, as it demands evidence of substantial similarity, not merely paraphrasing. I realized it would be a futile effort since the true value lies in implementing an idea, not just possessing it. I have long believed that "ideas are a dime a dozen; only implementing an idea has value." Furthermore, Champy and Hammer's limited experience in creating I.T. processes would have significantly hindered their ability to make use of it effectively.

London

The new system would be deployed in London, England, Mexico City, Mexico, Johannesburg, South Africa, and Indianapolis. With this wide geographic spread, we would need to support operations 24 hours per day, seven days per week, 365 days per year. So, we designed and planned to do so. The company had local Directors in each of the foreign countries. They managed autonomously for the most part, reporting directly to Christel, as did I.

During our design stage, I had a battle with the fiercely independent Director of our London office, who wanted his own mainframe computer, application system, and programmers because he believed the European operation, including the London office, was substantially different from the rest of the world. I was convinced it was not. The argument was whether a single centralized system was a better design than a decentralized network of local systems. I met him for dinner in London, and we ended up having a loud, un-English-like argument in a posh English restaurant. It seemed like we were screaming at each other, but it sounded worse to me than it probably was.

Ultimately, I convinced our boss, Christel, that centralized was the best way to go and that his claim of being different was insufficiently founded. I would later be proved correct. All transactions were already conducted in U.S. currency, all the member rules were the same, and we designed the database to handle different formats for phone numbers, postal codes, etc.

My only concern was that the distance might be a challenge for our objective of sub-second response time from Indianapolis to London and South Africa, and back. Sub-second response time was measured from hitting an enter key to seeing the data needed on the screen. When talking to a member on the telephone, an agent must be able to find the member's information on the system quickly. Sub-second response sounds fast, but even a whole-second response time would be noticeable to a member waiting on the phone for an answer, especially if multiple screens needed to be accessed.

I was able to delegate many aspects of the development effort to my managers because I trusted them and because they were following the Development Process that we had evolved from my days with Barbara Allen at United Airlines. But not all issues fit into a smooth process.

Three of my managers approached me with a technical problem at one point in the development. They needed the mainframe to perform some function, I do not recall the exact problem, but they said that our MVS (IBM Operating System) expert told them it could not be done. Harking back to my days at Stat Tab when I did much experimenting with the internals of the IBM 1401 and 360, my gut told me that the IBM 360's could do what they asked. Our MVS expert was Jerry Harless (sadly, now gone), and his partner during the conversion project was Robert (Bob) Burns. Jerry was so knowledgeable about MVS that IBM often called him for advice. It was bold of me to call our expert's advice into question.

I invited Jerry to my office, talked about what my programming managers were asking, told him I was reasonably sure that the IBM360 could do it, and asked him to go back and look further. About 6 hours later, he returned to my office and admitted that I was correct that the IBM360 could do it; and he had so informed the programming managers. Never underestimate the lessons you learn early in one's career.

RCI de Mexico

Getting service into our Mexico City Office would also be an issue. We arranged the dedicated data lines from Mexico City to the Texas border and from Indianapolis to the Texas border. However, neither Telecom de Mexico nor AT&T would accept responsibility for connecting the two pairs of lines from each side. As a result, I had to dispatch one of my technicians to the Texas border, where the two companies watched as he spliced the wires together.

The other challenge was that IBM could not import PCs into Mexico. Mexico was holding out to try to force IBM to build PCs in their country. Negotiations were underway, but we needed to get one IBM PC into the office so they could help test the system and begin utilizing it. My boss, Christel, planned a trip to Mexico and invited me and our CFO. I decided to take an IBM PC with me. Technically, I would be smuggling it into the country. A manager working in the Mexico office had an uncle at the time who worked in the customs area of the Mexico City airport. We were told the uncle would meet us at customs.

We connected flights in Dallas, and when we were boarding the connecting flight, I was shocked to run into two US Customs officers in the jetway as we were boarding. Having traveled a lot, I had never encountered Customs inspecting passengers leaving the country.

One of them asked me what I was carrying. I told him I was bringing a PC down to our Mexico City Office to give a presentation, which was partly true. I just did not plan to bring it back, but he did not need to know that. Puzzled, he flipped through papers looking for whether my contraband was legal. At the time, it was illegal to export advanced electronics, and the PC was new and probably qualified as advanced. Another more senior agent arrived, and together they puzzled for a while. Finally, the senior agent gave the first agent an unknown reference number that he could use and told him to let us proceed. Whew! I passed that hurdle.

When we arrived in Mexico City and entered the Customs area, my boss, Christel, got extremely nervous about my contraband. She would have run back on the plane if she could. As we approached the area where bags were inspected, we saw a tall and massive man flanked by two customs agents walking briskly toward us. I was not nervous, mainly because he smiled as he approached us. Christel was muttering something under her breath, probably that she would kill me. However, it turned out that the big man was the uncle of the manager in our office. He welcomed us graciously and escorted us straight through customs without inspections.

Our local director, Gabriel Oropeza, a brilliant and personable young man, greeted us outside customs and showed us to a waiting car. We loaded our luggage into the trunk. My boss and I got into the back seat, with me putting the PC on the seat between us. Our CFO sat in the front seat with our local director. We started across the city towards our office. After driving for about 30 minutes, our director suddenly pulled the car over and stopped. He said that there were police behind us and not traffic police but rather Federal Police. At that moment, I became extremely nervous. Had someone turned us in? Would I be arrested and end up in a Mexican jail? Would my boss be arrested and jailed? I could do nothing but wait, and we waited for some time. We heard the trunk being opened.

Finally, our director returned, and as he got into the car, he explained that the police had noticed that the car was riding low and wondered if we were transporting illegal drugs. Thankfully, they never looked in the back seat to see the large package I was leaning on that could easily have contained drugs. We got the PC to the office safely, and I was able to set it up for them. They could start becoming familiar with desktop computers.

This was also when I first recognized the power of systems as a driving force in marketing. By building an impressive system in a striking building in a showcase computer room, Timeshare Developers would be impressed and choose RCI over our competitor, Interval International (II).

Top 200 Industry Salespeople

Once the system was implemented and became well known, RCI held its annual gathering to recognize the top 200 salespeople from the Industry's RCI-affiliated Developers.

The company had surveyed the attendees ahead of time and asked them what subjects they would like to know more about. Naturally, the new Exchange System was a hot topic, and I was asked to give a one-hour talk on the system and technology. With my experience at Stat Tab and United, presenting to senior executives, I was confident I could give such a talk. However, before I sat down to plan what to say, I decided to research how to talk to industry salespeople.

Our Marketing VP, Herb Alfree, had decades of sales experience in the field, so I sat with him to pick his brain. He gave me lots of pointers that I took to heart. The main one was to avoid boring them since they had short attention spans. My talk had to be fast-paced and varied. So, I included such things in my talk as

1. Helping them visualize what a megabyte of data might look like by leaning on a stack of 32 reams of

copy paper (16,000 sheets) and relating that to 16K of memory and that to one megabyte of storage, which would be the equivalent of 2,000 reams or 63 stacks like the one I was leaning on.

2. In the middle of my talk, I paused while a couple of my programmers passed out Godiva chocolate bars to everyone in the room while I made a point about the importance of quality in everything we did.

3. At the end, I shared an anecdote about Jon DeHaan, a successful stock market investor who claimed he did not want to be a Bull or a Bear but rather a Pig. At that moment, I pulled a pink balloon with ears, a snout, and a curly tail from a brown paper bag. This got me a final big laugh.

They gave me a standing ovation. This was my lesson that to be successful with large data processing projects, you not only have to do them right, but you also must sell them well.

Risk Management

Any time you take on a significant endeavor, you take risks. Risks require trust in your team and self-confidence, which I never thought of myself as having but must have. I faced several risks in championing the complete rewrite of the company's core operating system. The usual risks of designing and implementing a system that might not work, building a system with too much or too tightly packed functionality making the system too processor-heavy even for our large 360 computers, or having a new system that the company relied on that is full of bugs, damaging the company and frustrating the Operations department.

I had to trust my programming managers and our process. If I ever had a slogan, it would be to "Trust the Process," which assumes you have a solid one. We hired Arthur Anderson to review our development process and compared theirs to the one I had first developed at United Airlines and then refined over time. We found a few gaps in both, but by consolidating them, my team built a new methodology incorporating the best of both. And it did the job on this project.

To mitigate risks, one option is to buy or license an existing system developed by someone else. In our search for such a system, we investigated one operated by Resort Computing Corporation (RCC) in Denver, CO. I met with the owner, Dennis Torgerson, multiple times to discuss acquiring or licensing his system, either to integrate it into RCI's Exchange System or to offer it as an additional product. Although Dennis was forthcoming and provided a detailed flowchart of his entire system, we decided it was not a good fit with RCI's core business and passed on it. Later, I considered it again for potential acquisition when I was working on the systems for Disney Vacation Club, but decided against it there, too, because it was primarily designed for a single local resort. RCI eventually acquired the system but did little with it.

Hiring risks were part of this project too and throughout my work life. We had to hire the best programmers, trainers, and procedure writers we could find. One risk was hiring a college student who worked in the mail room, pushing a cart around the offices to distribute mail to people's desks. One day he stopped me and told me he was studying computers in college and would like to become a programmer. Being sympathetic, I hired him after he took his first few programming courses. He became one of our best programmers. His name is Dave Schacht, and after our project, graduated and later served for 23 years as SVP and CIO of Simon Property Group, the Mall Developers and Operators, and still works as a consultant. One of my greatest accomplishments!

As I mentioned earlier, I also oversaw the warehouse and mailing operations. A Vietnam couple worked there after being sponsored by the DeHaans as refugees from the group of immigrants known as the "Boat People." Many of the Boat People escaping from Vietnam at the end of the war traveled in overcrowded and unseaworthy vessels, often facing danger, piracy, and starvation during their journeys. Some estimates suggest that as many as one-third of the Boat People perished at sea due to these dangers. I heard ours suffered horrendously while escaping.

The husband, Ken, came to me one day asking how he could get into the computer side of the business to support his family better. I knew that because of his heavy Vietnamese accent, many people in the company had trouble understanding him when he spoke, which limited his options in the IT department. Until one day, I had an idea. He could learn to be a computer operator. That job did not require much communication with users and was eminently learnable on the job. I helped him transition into computer operations. Then, to enhance his resume, he began attending computer classes in the evenings while working as a computer operator during the day. He had endless energy.

His family was forever grateful and showed their appreciation by occasionally bringing my wife and me large trays of Vietnamese eggrolls. I continued to mentor him and once advised him against a potential entrepreneurial investment idea he had to sell imported surgical gloves. I cautioned him against putting his hard-earned money into a business where he would compete against large, entrenched companies that could easily undercut his pricing and leave him with nothing. He later realized that I was right and that I had saved him from a potential financial loss. Although I'm unsure where he ended up in his career, I knew he was a hard worker, driven to succeed, and a survivor who would ultimately be okay.

Taking risks with people has always worked out well for me. I have always been repaid with loyalty, hard work, appreciation, and a few egg rolls.

But these were not the biggest of the risks I took. The most significant risk was whether the data lines could deliver sub-second response time globally. This was a vast unknown – how would our data be routed across the Atlantic? Would our dedicated lines pass through London and then down the length of Africa? What about through Mexico? There were no satellites yet; every line was hard-wired or fiber-optics. We had no idea but had to trust AT&T, Telefono de Mexico, and other local telephone companies in foreign countries.

RCI-ACD

I again had to acquire an Automatic Call Distributor (ACD) for the new offices to manage the rapidly growing volume of inbound calls. I again went with the Aspect ACD because they were flexible in tailoring it to our needs, and the system was far less complicated than the Rockwell-Collins ACD, which was still the standard for many large call centers.

ABCD

Managing such a large I.T. team made recognition especially important. And I always believed in including everyone in goal setting and recognition. My favorite way of recognizing programmers was through the ABCD Award, an idea originally suggested by one of my managers but enhanced by team input.

ABCD was shorthand for Above and Beyond the Call of Duty. The main rule was that anyone could nominate anyone, even themselves, when they did something well beyond their normal duties. The nominating person had to submit the reason on a form signed by them. I was the ultimate judge, albeit a generous one. If approved, the nominee would receive a check for $50 to take their significant other to dinner to celebrate. The awardee also received a certificate describing how the person went above and beyond the call of duty, who nominated them, and my approval signature.

The program became highly popular, and I only recall turning down one recommendation out of hundreds. That was only because the nominator withdrew the award after deciding that the recognized task was part of the nominee's regular duties. I enjoyed walking around the office and seeing all the certificates pinned to people's walls.

Anti-Smoking Campaign

To show how much the company cared about its Associates, I got the company to build a fitness center on the ground floor, with room for aerobics classes.

One of my managers went to a local fitness center to interview a potential aerobics instructor to work part-time in our company fitness center. She had invited us over to take a beginner's class. We joined in but quickly became exhausted and had to bow out. After the class, I asked, "That was a beginner's class?" She responded, "Oh no, sorry, that was an advanced class." I felt better but tired.

I initiated a campaign to encourage our associates to quit smoking. Out of our 120 team members, roughly one-third were smokers. I offered a reward of $200 to anyone who could quit smoking for two months, figuring that was long enough to break the habit. Many from my team accepted this challenge. We implemented an honor system where individuals who relapsed during their two months had to reimburse the company. Some did end up paying the company back. Despite this, I persisted in urging everyone to stop smoking and began promoting the objective of having everyone quit. It became a team culture thing and an expression about how much the company cared about their health. At one point, for a single day, everyone agreed to abstain from smoking. The next day, however, two members resumed smoking, but about 40 associates remained smoke-free permanently. This stands as one of my most satisfying achievements.

RRP

Around this time, the industry association American Resort Development Association (ARDA), under the leadership of Perry Snyderman, was bent on cleaning up the poor reputation of the timeshare industry – a reputation for high pressure, shady dealings, and even scams. Perry created an Ethics certification program called the Registered Resort Professional. I immediately jumped on this because I believed strongly in his program. I was the first to obtain the certification at RCI and one of the first in the industry. After almost 40 years, this certification is still a standard for the industry.

Cutover

After eighteen months of demanding work, the Indianapolis cutover was scheduled to go live over the next upcoming weekend. Our cutover process called for increasing levels of "volume testing" leading up to the big weekend. The first two volume tests had succeeded with flying colors, and my management team came to me and asked me to cancel the last one, scheduled for Friday, because they had too much to do to prepare for the weekend. The three scheduled tests were at increasingly heavy volumes. I considered their request for a minute or so while they stood in front of me. I was sympathetic but told them we had a process for a reason, and they still had to do that third volume test.

To their surprise and mine, the last day's test failed. The entire department was immediately depressed. I had to take the whole department outside to a grassy area behind the building and talk to them. I told them that the failed test did not reflect a failure of their work on the system. The test was a planned part of our process, and our process successfully prevented us from messing up the company, which meant it was successful. I told them we would take two weeks to sort out the cause of the problem and cutover on that weekend. Fortunately, the problem was solved in two weeks, and the cutover went flawlessly.

South Africa

We did not cutover South Africa during the first week. A few weeks later, I traveled to Johannesburg to oversee turning up their terminals. The flight to South Africa was grueling - flying overnight to London, arriving in the morning, going to our London office to ensure they were up and running, then heading back to the airport for another overnight flight to Johannesburg. Once I arrived, I went straight to their office to see their response time for myself. Despite the significant risk I took in committing to sub-second response time at the project's outset, I was pleasantly surprised to find that their response time was both sub-second and virtually instantaneous. As a result, we were able to provide sub-second response time worldwide, likely for the first time for any network for any company anywhere.

At the end of the workday in Johannesburg, we flew to the east coast of South Africa. After 72 hours of non-stop flying, visiting offices, and only managing snatches of sleep in-flight, I finally had the opportunity to spend my first night in a bed.

We toured their resorts the next day and took off again in the mid-afternoon.

Near Death Experience(s)

I had a sobering experience on this trip. I flew with the local company Director to Pretoria on the east coast, so I could see one of their vacation areas. The Director, a manager from his office and I flew on a small two-engine aircraft. On the return from Pretoria, because it was getting late, they had arranged for us to drop into a game reserve to stay the night. As we approached the area, the Director and the pilot argued about where the pilot was supposed to land. Finally, the pilot made the decision and landed on a muddy dirt strip inside the game preserve. There was nothing around us once we disembarked except a damp dirt road along the edge of the strip.

I asked how we would get to our lodging, and the pilot said one of us would need to walk down the road to a house about a mile away. Having previously visited East Africa, which boasts similar wildlife to South Africa, I had a sense of the behaviors of their wild animals, their environment, and the risks associated with walking on foot through a game area. I refused on behalf of myself and the two guys from our office, telling the pilot that he would be walking alone if he decided to go down the road. A memory flashed back to me: I had jokingly shared with my wife on multiple occasions that, when I die, I would like to die in an interesting way, "like being eaten by a lion." So, I knew that I was not going to tempt fate.

Soon a land rover came by on the road. The pilot asked the driver to stop at the house down the road and have them call our lodge to come and pick us up. He did, and soon we were picked up and driven to the lodge. As soon as we were settled, the lodge staff took us on a "game drive." I bought a hat in their miniscule gift shop. We rode in what I would describe as a World War II U.S. Army jeep—on steroids—a six-passenger version open on the sides with no top.

After searching for game without success, our guide and driver received a call from another guide telling us to drive to where he was parked near a large male lion. We parked within 20 feet of the lion, who appeared alert but was looking away. Nothing stood between me and him except air. I asked our guide why the lion did not attack us, and he explained that the lion perceived us as part of the vehicle. The guide warned me that stepping out onto the ground would result in a quick death. As this was too close to my previous joking remark about wanting to die in an interesting way, I opted to keep my feet inside the vehicle.

Soon, we heard noises from the bush in front of us. Our guide and driver said the females were in the bush, hunting. Suddenly, the male lion stood up and ran around to our left. Within moments, a dozen antelopes appeared directly in front of us and ran past having been flushed out by the female lions then emerged from the bush, joining the male on our left. They then walked toward us and then across in front of us. When the front of this parade was passing, we saw a young cub jump into the grass and lay down, appearing to be preparing to ambush the male lion from cover.

Our guide spoke about how difficult life can be for male lions and how they lack any "sense of humor." As he backed the vehicle away, he warned us that we might witness the cub being killed right in front of us. When the male lion passed by the cub's hiding spot, the cub jumped out and swatted the male on his rear. The male lion spun around in a flash and swatted the cub, sending it rolling in the dust but without hurting it. Our guide informed us that it was something that he had never seen before and that none of us would likely ever it see again. He added that that male lion was the most confident lion he had ever witnessed.

On the way back to the lodge, we were caught in a deluge, a rainstorm so heavy that I could not see anything beyond the rim of my hat. It was the heaviest I have ever experienced, and we got soaked. I do not know how our guide found the lodge, but he did.

That night, after changing into dry clothes while dining on exotic game meat and other local foods, heavy rain continued to pour down on the lodge. We learned from the Preserve Owners that the heavy rains from the previous night had washed out a bridge upriver from the lodge, which explained why our landing spot had been unsafe.

That night, I slept in a private hut, but my rest was restless due to sleeping under mosquito netting, the hut having only a screen door separating me from wild animals, and the nearby river that could be cresting its banks.

The next morning, the pilot informed us that the landing strip was even muddier. He recommended that the lodge transport us and our luggage to another airstrip. In the meantime, he would attempt to fly the plane alone, without any passengers or luggage, from the muddy strip where we had landed and meet us at the new location.

When we got to the other airstrip, I was surprised to see it was an actual small airport with an asphalt runway and a small terminal building. I knew then that the pilot had made an unwise decision the day before to land where we did.

As we arrived, I could see thick clouds descending on the airport. We would have only minutes to get away before it was socked in. As we eagerly watched, we saw our pilot emerge from the thick cloud cover no more than 20 feet above the runway. As the plane touched down, the clouds were swirling behind it. As soon as the pilot landed, we rushed our bags and bodies into the plane, and he took off as visibility disappeared.

As the plane labored upward, I could hear the pulsing drone of the engines struggling for altitude. The rain was deafening as it pounded on the windshield. I was in the right front seat, studying all the gauges trying to decipher them. I wondered how the pilot could tell up from down with zero visibility outside. Before long, while still in a steep assent, or so I hoped, a cockpit alarm started screaming. The pilot (and I) anxiously studied all the instruments to identify the problem and see if we were in serious trouble. Then, the thought crossed my mind, if we crashed in the game preserve, I might very well be eaten by a lion!

The pilot reached over to a small panel in front of me, popped it open, and started pulling out fuses and pushing them back in – until the alarm stopped. I asked him what he had just fixed, and he said he disabled the alarm. He then clarified that it was an alarm to warn about "stalling." Stalling occurs when an airplane's wings are tilted too far upward, causing the wind to stop flowing over the wings, which holds an airplane up.

I could not tell if the plane was falling or still rising, that is until we broke out of our cloud into a brief, clear spot where I could see we were still rising. The pilot said we were okay because he could tell by the feel of the controls, but I still was not sure if my death wish might still come true.

As soon as we landed in Johannesburg, I promptly boarded a commercial flight to Indianapolis, with a layover at Heathrow airport in London, enduring another twenty-four hours in coach class.

Another Near-Death Experience

The computer room became the company's centerpiece for hundreds of timeshare developer tours in the months after the system was up and running. The new facilities and its state-of-the-art system significantly elevated the company's reputation. It became a central selling point for RCI's developer sales team.

The company made a significant push at the next ARDA industry conference in Miami by inviting over a hundred industry developers to visit our offices in Indianapolis at RCI's expense. The company chartered an American Trans Air DC10-10 out of Indianapolis, IN, to fly the developers up to Indianapolis for a tour of the new office and back to the conference again. That was a bold move by the company's owners.

When the plane was set to leave Miami, because I had often flown in the cockpit at United Airlines and knew that pilots had the prerogative to allow me to do so, I asked the captain for permission to ride in the cockpit, which he granted.

(Note: This is no longer allowed after 9/11/2001). The DC10-10 was a large airplane, seating 380 passengers, and it departed full.

Upon the flight's approach into Indianapolis, I was wearing a headset and listening to the tower. There were four of us in the cockpit, the pilot, the co-pilot, a navigator (a position no longer needed on the DC10 but carried over from earlier days), and me. My seat was behind and raised about six inches higher than the pilot's. The navigator was to my right, behind the co-pilot, and situated slightly lower.

The co-pilot was in control, and through earlier conversations, I knew he had been an Alaskan Bush Pilot in an earlier life. I heard on my headset when the tower warned our pilots that a small aircraft was in our vicinity and to be alert. The tower had tried to radio them, but the small plane's pilot was not responding. This alert put all four of us in the cockpit on high alert, looking in all directions.

In an instant, the captain in the left seat was looking left, the navigator was too low to see anywhere but upwards, and the co-pilot in the right seat was looking right. Suddenly, I saw the other plane appear right in front of us. I immediately yelled and pointed to alert the crew. The co-pilot, who had lightning-quick reflexes from his bush pilot days, took swift evasive action, diving our airplane down and to the right just before the small plane passed overhead. We had just experienced a "near-miss," or as George Carlin liked to joke, a "near-hit." But this was no joke; it was a major near accident.

If I had not yelled at the instant I did and the co-pilot had not acted so quickly, the entire group of timeshare industry leaders on the plane (including myself) could have been annihilated. Fortunately, we will never know what could have happened. I can only say that the captain handled the situation professionally when he reported the incident to the tower and requested that they track down the person flying the other plane and ensure they were grounded. After his call, he expressed his frustration with a few expletives and described what he would do if he could get his hands on the other pilot.

I never told any of the passengers what had happened, and none of them distinguished the sudden aircraft movement from normal air turbulence. I also found it ironic that the pilot refused to let me fly in the cockpit on the return flight. Imagine if that pilot had been the one on the flight up.

Yet Another Near-Death Experience

Who would ever think being head of IT for a timeshare Exchange Company could be dangerous? I sure did not. On a trip to meet the Director of our Mexico office, my boss, Christel, and I just arrived at the Director's office. We had greeted him and were ready to go to lunch when he was interrupted by his Assistant for an urgent call. We stood by his desk as he spoke on the phone. He said, "Uh-huh, uh-huh, uh-huh, ok." When he hung up, he said we were invited to lunch, and attendance was mandatory. Puzzled, we followed him to his car.

He pulled up in front of a restaurant that appeared closed. The only other car parked out front was a black Mercedes with darkly tinted windows. We saw an older gentleman and a younger man sitting at a table in a small private area near the entrance as we entered the restaurant. The rest of the restaurant was empty and dark. The man introduced himself and his son. We were served lunch, but I do not recall eating anything.

Next, the man asked Christel to explain her business, which she did. He seemed pleasant enough. He then asked her about a lawsuit the company had filed against a resort in Cancun. She explained how the resort had signed a contract and then made another agreement with our competitor, and she was bound to hold them to our contract.

The man was silent for a few moments, then he asked her to withdraw the lawsuit and that he would see to it that the original contract was honored. He then said that he would "take care of things the Mexican way" if the word of his agreement with Christel got back to the resort's management. Then he formed his index finger and thumb into the shape of a gun, pointed it toward his temple to emphasize his point, and then pointed it towards each of the three of us and said, "You. And you. And you," while moving his thumb like the hammer of a pistol. One of those threats was pointed directly at me, and I admit it sent shivers down my spine.

After we left, we all agreed to keep the discussion to ourselves. I only speak of it now because it happened decades ago; the man was old enough then that he, indeed, must have passed on by now, Christel has passed, and no one could tell from this story which company, which resort, or which resort manager was the topic of our conversation. This experience was my first direct death threat, but not my last.

Post RCI Exchange System Projects

After the big projects, I helped our VP of Marketing, Herb Alfree, arrange for the company to acquire a small Travel Agency company based in Grand Rapids, MI. It was called Endless Vacation Travel (EVT). The business was an excellent complement to RCI's timeshare exchange business. It would grow with RCI's added business to become the country's first $ 100 million travel agency. Their systems were standard travel agency systems and required no integration with the Exchange System, although we did enable call transferring from RCI's agents to EVT's agents to facilitate member travel arrangements when requested.

At Herb's suggestion, I analyzed unused exchange inventory by destination and season, which we named "Breakage Weeks," meaning weeks that no Exchangers had selected, which RCI had the right to utilize and would go unused and expire in our system. I created a certificate with rules and limitations that developers could buy and use to entice families to visit their properties and purchase timeshares. Buyers could trade these certificates for free vacation weeks with only a transaction fee. Developers were our target market.

We sold certificates providing access to weeks in mostly shoulder and off-season periods at $129 each, usable for one week of vacation. We sold tens of thousands over the first year following the product launch, generating profit almost entirely. We also charged the timeshare buyer an exchange fee of $69 upon redemption. I estimated the company added $10-15 million annually to its bottom line from this new product. Our model became an industry standard, with our competitor matching us and some developers using their resorts' surplus inventory this way. Overall, the project required extensive analysis, modeling, and data processing.

During my last two years with the company, the husband-and-wife owners decided to divorce. However, during the six years I worked with the five VPs, we substantially improved the company's profitability as a team. By my last year with them (1990), the company made $65 million in profit, a significant improvement from the loss they had the year before we started. Despite the disruptive and contentious divorce proceedings, the couple eventually settled, with Jon receiving $650 million in taxable cash and Christel owning the business. Later, Jon was awarded an additional $135 million to cover his taxes. Christel eventually sold the company for approximately $1 billion, a remarkable achievement for the couple, who had started the business in their garage with just some 3x5 cards. They each pocketed almost a billion dollars from the sale, but unfortunately, none of us VPs received any part of this enormous cash-out.

After working for six years, I realized that the company's opportunities for further meaningful change had been exhausted, and I started exploring other career prospects. Unfortunately, Jon distrusted technology due to his lack of understanding, so he often sent an Arthur Andersen consultant to try to supervise my work. This consultant, whom I will call Bill M., was unfamiliar with building large systems and needed to gain an appreciation for the programming talent that one gains from having been a programmer. Instead, he was a "glad-hander" who could talk a good game but lacked the skills to perform the tasks required for my job. His occasional oversight of my activities was a source of frustration for me.

One thing I learned about Bill M. was that he had no empathy. I even told him as much at dinner once, and his response was telling: "What is empathy? I do not know that word." Sadly, this was all too true, as he never showed understanding or compassion for others.

As soon as Christel took over the CEO reins, she hired Bill M. and placed him over my position in preparation for letting me go, I figured. The handwriting was on the wall, as the saying goes. Meanwhile, I interviewed at Disney and was offered a position as Director of IT for their Disney Development Company (DDC). The director role at Disney is equivalent to the position of VP in most other companies, even qualifying for first-class travel and invitations to grand openings of new films. DDC had plans to investigate the timeshare business but needed someone to solve several problems in the Development Division immediately.

I arranged a meeting with Bill M., my new RCI boss, planning to inform him that I was leaving the company. When I entered his office and started to speak, he interrupted me and said he had something to say first. He then informed me that they were "letting me go" and presented me with a six-figure severance check. I thanked him and left the office feeling surprised and very pleased. I did worry about how my departure would affect my team, but I knew there was nothing more I could do for them at that point.

The following Monday, I reported to DDC, and my wife and I were put up in a company-provided condo. We learned about how popular Disney World was because visitors from up north stayed with us for five of the eight weeks that we lived in that small condo. We soon had to find a house with guest bedrooms.

What I learned at RCI

As a technical person moves from being a programmer, through systems analysis, into management and up the chain of command, they start to lose their technical skills because managing people and keeping programming skills current is challenging. When transitioning from a technology-focused company to a non-technical organization with limited technical requirements, your proficiency in technical skills may diminish unless you possess extensive experience and actively work to maintain and enhance those skills in your free time. I have continued to develop and uphold my technical expertise throughout my career.

I regularly sought innovative ways to motivate my large teams to maintain their enthusiasm for technical work. One effective method I employed was fostering a sense of camaraderie through friendly competition among the Information Technology (IT) Department members. For four successful years, I organized an Annual Chili Contest that brought the team together and encouraged their engagement and participation in a fun activity.

Anyone in the IT Department could enter, and other departments could submit one entry. Three judges were selected -- two managers from non-IT departments and one outside "celebrity chef." The only rule was that there were no rules, although that was not entirely true.

Unofficially, I told my IT department managers that they were eligible to enter but would not be eligible to win. They joined in for fun, so I told the judges that the managers were not allowed to win. In each contest, my managers would loudly and jokingly protest to the participants that their chili was the best and that they should have won if it was not for the judges being biased against managers. Their involvement became a running joke between them and me, especially since I suspected the managers were buying their chili at a chili parlor in town. I entered, too, with an original Mexican version. My chili included a couple of sticks of Mexican cinnamon for extra flavoring, and one year; I forgot to remove them. The managers bought a trophy and awarded it to me for having the "Most Miscellaneous Debris."

For the fourth year, I thought I would outsmart my sneaky managers. They had never admitted to buying chili from a professional chef, but I heard a scuttlebutt confirming it. So, I went to the chili parlor where I believed they were buying their chili and invited the chef's wife and partner to be our celebrity judge. I hoped to shock the managers at the start of the judging when I announced who the celebrity chef judge was. However, it turned out even better.

When the managers went to buy their chili, the husband unwittingly told them about his wife being the judge. That put the managers into a panic. They were convinced that not only did I have the judges set against them, but there must also be a bigger surprise in store for them. They sweated bullets, worried about what else might be coming until the judges were announced. Their panic was even more satisfying than the simple surprise I had planned.

Firsts at RCI

My team, fellow associates, bosses, and I achieved several firsts at RCI:

1. We created and implemented the first global online real-time computer system.
2. We developed and implemented the first online system with a sub-second response time halfway across the globe, and it could have spanned the entire world if RCI had had offices in the far east.
3. We installed the first telecommunication lines across the U.S. – Mexico border.
4. We delivered the first IBM PC into Mexico.
5. AT&T invited me to sponsor their legal action against the anti-trust regulations that had kept them from competing against MCI and other telecommunications companies. The other companies could underbid AT&T, and AT&T could not match them as they were forced to stick with rates published in FCC Tariff filings. They asked me to be the complaining party in their legal action because RCI was their largest customer in the Midwest. They won the ground-breaking suit, and my name is on the cover of that ruling somewhere. It also allowed them to win RCI's business.
6. Herb Alfree and I developed the first vacation certificate in the industry.
7. I learned how to prepare a Request for Proposal (RFP) and manage the processes around doing an RFP.
8. I implemented the first Above and Beyond the Call of Duty (ABCD) Award program with my team, which I would replicate in future jobs.
9. I encouraged about 40 people to quit smoking.

CHAPTER 5

DISNEY DEVELOPMENT COMPANY (DDC)

During my first week at DDC, I missed three meetings, unaccustomed to their consensus-based decision-making approach. In my past roles, the decision-making process relied on clear delegation, granting signing authorities, and fostering independence. These elements formed the foundation for effective and efficient decision-making within the organizations for which I worked.

I have always preferred working in organizations where responsibilities are clearly defined, goals are agreed upon, plans are set, and individuals work as needed with others while pursuing their initiatives. Managers who have signing authorities appropriate to their position can capitalize on opportunities and make prompt decisions. This delegation fosters fiscal responsibility and accountability for their actions. Initially, the DDC environment felt unfamiliar to me as I was used to making decisions independently and attending fewer meetings. However, I adapted to the new environment by the second week.

My first tasks at DDC were to standardize their network of desktop computers and to set up a JD Edwards automated accounting system on AS400s for both the Orlando office and Euro Disney. The development of Euro Disney, later known as Disneyland Paris, was about to begin.

DDC mainly comprised architects, accountants, and engineers, with half using IBM PCs and the other half using

Apple Macs. Both groups were passionate about their chosen brand and unwilling to switch to the other brand.

My boss, Matt Ouimet, who later would be Chairman of Cedar Fair Entertainment Company after 17 years at Disney, tasked me with solving this problem, which was mainly political. Matt did not want to be the bad guy who made half the staff change desktops. And I realized that either way, more than 50 associates would need to be retrained, and some of the software they relied on likely would not transfer if we forced everyone to one brand. Unfortunately, at the time, these two devices would not network together. IBMs would network using Ethernet connections, but Apple MACs would not.

As soon as I formed a team, I assigned my technical manager, Jim Reichley, to find a way to network IBM PCs with Apple Macs. Two weeks later, while I was on a business trip to Los Angeles, Jim called me to let me know he had located a software firm in Phoenix developing an Ethernet connection specifically for Macs. I requested Jim provide me with the firm's address and let them know I would visit them on my way back from Los Angeles.

When I got to their office, I learned that the product was only an early Alpha test, not even in the Beta test stage. A software alpha test is the first phase of software testing and is usually conducted internally within a company or by a narrow set of external users. On the other hand, a software beta test is the second phase of software testing and is conducted by a group of external users who use the software in a real-world environment but in ways that do not expose the testing company to undue risk.

I asked the company if they would let me take a copy to try it, promising them feedback on how well it worked for us. They agreed but needed Apple's approval, which I requested them to obtain. Unfortunately, when they tried to reach their contacts, they found that all their Apple contacts were out of the office at a team-building event. I located the corporate phone number and contacted Apple headquarters, hoping to speak with a management representative. Fortunately, I connected with a young company attorney who was one of the few not attending the team-building event. I explained our urgent need to her and promised to return the software if she received different instructions once everyone returned to the office. She agreed and permitted the software company to release the software to me. I returned to our office with the software on a floppy disk.

Jim Reichley had the software installed and operating the next day. We had avoided a major uprising and had gotten the job done. In a few weeks, after I worked with my boss to prepare a set of General Ledger accounts, Jim installed the J.D. Edwards Accounting Software for the DDC headquarters operation.

DDC- Euro Disney

My next challenge was to build and implement a PC network and J.D. Edwards accounting system at Euro Disney, as we had in Orlando. I encountered several challenges in Marne le Vallee, where Euro Disney was to be constructed. Firstly, there was no existing technical infrastructure in place, and secondly, I needed a technical manager like Jim to provide ongoing support. Third, the language barrier was a significant challenge as I did not speak French, and in the region of France where the DDC offices were, few spoke English.

I arrived at the office building where the park development team was working on creating the theme park. All that I had been given was an address -- no contact person. When I arrived, I found one person who spoke a little English and asked who oversaw Euro Disney.

I was told Philippe Bourguignon, Chairman and CEO of Euro Disney, was the lead executive. I managed to find his office. When I got to his office, his very kind Administrative Assistant, whose name I have long forgotten, spoke with him on my behalf. He instructed her to set me up in a small office outside his office. It was furnished with only a desk, chair, and telephone. That was it. I felt more than a little "at sea." I was determined to get started, and I prioritized hiring a bilingual local manager.

With the help of his assistant, who spoke no English but did well with hand gestures, I met with the local HR person. That person knew some English. I provided her with the job skills for the on-site IT manager, equivalent to Jim Reichley, that I was seeking. She agreed to arrange interviews with candidates for my next visit. I planned to travel to France for one week every month for as long as needed.

On the second trip, I interviewed the candidates. The most qualified was a young professional woman with solid technical training. I hired her without realizing how women in France were treated in management roles. At that time, Frenchmen generally treated women like second-class citizens. However, I was genuinely fortunate because the lady

I hired took no guff from any of the men and soon established herself as a respected team member, especially after she set up the AS400, Apple Mac Ethernet network, and J.D. Edwards accounting system in six weeks.

While working with other Disney people there, I talked to the person who had built the analysis that decided whether to locate Euro Disney in France or Spain on the Costa del Sol. He let me see the actual spreadsheet. While looking it over, I noticed a significant flaw in the analysis. They used the same seasonality for both locations, when the seasonality in Spain was much better, with more extended periods of warm weather and less winter. If they had corrected that assumption, the park would have been built in Spain. The young man who had created it acknowledged the flaw, and as it was too late to change locations, I agreed not to share what I discovered.

Also, on one of my early trips to the park site, although it was not my job, I was concerned that the Disney concept of a theme park might not be as widely accepted as the planners believed. The French children might not relate to Mickey Mouse. My local manager drove me to a small Visitors Center that had been built near where the entrance to the park would soon be. At the time, this was the only building at the park.

When we entered, we were escorted into a small auditorium and told that Mickey Mouse would soon appear. A busload of schoolchildren arrived and were seated in the front rows. As soon as they sat, Mickey Mouse appeared behind a curtain, and the kids went wild, cheering and gathering around. That experience put my concerns to rest. I knew the park would be very popular. Unfortunately, it would not be the financial success Disney had planned, and I am convinced that one of its weaknesses was the limited length of the "high season."

I also gave some input to our on-site designers, encouraging them to string lights over the walkways so that the park could operate when it was dark or overcast, thereby extending the season and the hours of each day. They took my advice.

On one of my early trips, I met with the local Apple MAC division in Paris once I had adapted to the bi-lingual environment. I had already met with their counterpart in Orlando, but that sales contact had yet to brief them on why I would be calling on them. I was there to tell them that Disney was counting on Apple to provide forty MACs at no cost to the Euro Disney Development Division. Disney had this kind of sway over its "strategic partners." When I told them this, they were shocked and insisted it could not and would not be done. I gave them the phone number and name of the U.S. person who had approved it. I promised them we would meet again on my next trip.

When I met with them the next time, they asked me to confirm that I was asking for forty free Macs, and I repeated my request. They again resisted but acquiesced when I insisted that Apple had already agreed. They may have been told to try negotiating with me but to yield if I insisted. So that transaction took care of our primary technological needs. My teams in both countries continued to work until the Euro Disney DDC team was electronically integrated with the headquarters' team.

I was fascinated watching the park rise from the ground in my once-a-month snapshots. I wish I had photographed what

I saw. On my second visit, my local manager took me to the field where the park and our hotels would be built. Huge earth-moving dump trucks were driving everywhere, each with tires taller than my height. On another trip, I counted over a hundred tall cranes. I heard Disney had hired every European tradesman to work on constructing the park. I watched the park grow like a sunflower, from seed to full bloom. I saw everything except the opening, to which I regrettably was not invited, probably because the Development Company was only building the five hotels and not at the forefront of the park itself and because my system installation work was completed. My local manager could handle ongoing maintenance on her own.

DVD Firsts

We accomplished several firsts at DVD.

1. The first network of IBM PCs and Apple Macs.
2. The first implementation of the J.D. Edwards Accounting System operating on an AS400 in Europe.
3. The first integration of two J.D. Edwards systems across the Atlantic.
4. Equipping the DDC office in France with Apple computers for free utilizing the power and mechanisms of partner marketing.

CHAPTER 6
DISNEY VACATION CLUB

My next project, and the one for which I was primarily hired, was to prove the feasibility of Disney getting into Timeshare. Many in Disney Management thought Timeshare, with its negative reputation, would be bad for the brand. I could see the potential but wondered if it would ever be internally approved.

We began as a new division of DDC with a separate small management team formed under Mark Pacala as president. Other notables were Cher Levine, Director of Member Services; Michael Burns, Director of Sales; Ralph Zeigler, CFO; and Bob Phillips, our head of accounting. I no doubt have forgotten the names of a few others. We planned to base the product on points. Mike Burns' father was the only U.S. timeshare developer whose product was based on points instead of weeks. There was one other in Europe, Hapimag.

We decided to call our product a "Club," which would be new to the timeshare industry and played off the original TV show, the Mickey Mouse Club. I recall our first meeting, which was focused on what the name should be. We settled on the name Disney Vacation Club (DVC) quickly. We began staffing and setting up our offices at the Westwood low-rise office complex in Orlando, FL. I recruited a small IT team whose primary responsibility was to create and implement the necessary systems to support various departments, including Member Services, Marketing, Sales, Reservations, Accounting, and the connecting network infrastructure. One of our first challenges was to flesh out the product's details.

One-Stop Shopping

The management team engaged in extensive debates on the Club's processes and unanimously agreed on the unique characteristic of "One Stop Shopping." This concept aimed to streamline the vacation booking experience by enabling members to accomplish all necessary tasks in a single call to Member Services.

Fulfilling this objective required recruiting agents with travel agency experience and developing a comprehensive system capable of handling all the various tasks. These tasks included tracking point usage, booking accommodations, billing, maintenance fee tracking, park ticket sales, restaurant reservations, and flight arrangements.

These system requirements were risky and would require integrating multiple computer systems and creating some systems that did not yet exist.

I brought Jim Reichley over from DDC with me, and we hired a couple of application planners and programmers for him to supervise. One of his earlier tasks, like the one I gave him to network the IBM PCs and Apple MACs, was to connect a pair of monitors to each desktop computer in Member Services. This setup would enable Member Services Agents to effectively simultaneously utilize up to four different systems displayed across their dual screens and seamlessly navigate the cursor across both screens, enhancing their efficiency and workflow.

Allowing multiple monitors is a common feature of operating systems and monitors today, but when we were creating DVC, this functionality did not exist until we achieved it, or if it did exist, it was not known to our desktop providers. Of course, Jim eventually figured out how to use multiple monitors on one desktop computer.

DVC Point System

To put what I was doing in perspective, you should understand that the entire industry was based on the sale and use of lodging in week increments. With only two small exceptions, over a thousand timeshare developers based their products on weekly increments. The Exchange Companies only worked using weekly increments. DVC was about to launch an entirely new approach where resorts could be booked in daily increments. This was one of the first "disruptors" -- before Uber disrupted Taxi services and before Amazon disrupted the retail business. Daily increments better fit what vacationers wanted and would give DVC a competitive advantage and make the DVC Member Program very appealing. It was my assignment to define the Point System and how that would work. I had two models to work from – Mike Burn's dad's (Bob Burns') system at Vacation Internationale in Bellevue, WA, which had a robust point system, and Hapimag in Steinhausen, Switzerland, whose points system was unknown to me.

Bob Burns was very forthcoming and told me more than I ever expected, including what he would do differently if he could start over. He certainly wanted to help his son, but I believe he would have helped us anyway; he was that nice of a person. On the other hand, Hapimag was not cooperative at all.

I traveled to Switzerland to learn as much as possible about Hapimag's point system. Hapimag was very guarded and reserved, although cultural differences may have played a part. Nevertheless, through close observations during my visit and reading their member rules, I deciphered how their system worked.

Issues of Substance

When my research was done, I constructed the Point Structure for the Disney Vacation Club. This necessity was a grand challenge because the point system had to integrate into the rules surrounding being a member. Our small management team had endless debates about how the rules should work, and we needed closure on the disagreements before I could complete the point system structure, which needed to be completed before we could begin designing and programming the systems.

So, one weekend I sat down and wrote a twelve-page document titled "Issues of Substance," in which I tried to decide how Members would use their points and make bookings. I then circulated the document to the management team, and the resulting explosion could have leveled Hoosier Dam. Everyone was yelling at me at the same time. However, in each instance of disagreement, I asked them to revise the document to fit what they felt it should say. I would redistribute the updated document to the team each time I received a revision. This process continued for many rounds, and we eventually ended up with a document that was 48 pages long. But then we had what we needed to program the supporting systems, and I had what I needed to finish the DVC Point Structure and design the supporting Systems.

In building the charts for different booking point requirements, as mentioned above, I had to consider factors such as seasonality, days of the week (weekdays vs. weekends), unit size, length of stay, and more. I even had to anticipate the potential for other DVC resorts to be built in different locations and the possibility of points from one resort being used at another. We had to factor in banking and borrowing, which are valuable benefits for Members. 'Banking' means deferring their points from the current year to the next, while 'borrowing' implies using next year's points in the current year, provided next year's dues are paid in advance. These are just a few examples of the complexities we had to consider.

I crunched many numbers and ran many simulations to sort out the possibilities. I also discovered an unexpected source of helpful information. Orlando Resorts and hotels in those days all had a "rate card" on their check-in counters. Every resort had already figured out when their demand periods were. Also, the Exchange Companies had done much work on the different demand scenarios for resort inventory. But none of these dealt with daily demand within a seasonal demand period. My math degree and statistical programming experience did serve me well during this creative effort. However, my biggest challenge with the points chart was getting legal approval.

DVC Final Approval

A two-day meeting was set up with Disney's legal department to review the point system, processes, and legal documents surrounding the registration of the product with the state regulatory board and what the contracts between the Members and DVC would say. All the legal documents were drafted, but the lawyers needed to review them with us. The meeting included 19 attorneys. Two of them were outside timeshare specialists who helped us prepare the documents, and two were personal attorneys of Michael Eisner, Disney's president, who were there to protect his interests.

Only two of our management team were present, Mark Pacala, our CEO, and me. The meeting started at 8 AM, and by the 10 AM break, Mark told me he could not take any more legalese and bowed out, telling me to "handle it." I stuck in for that very long day and the next. Ultimately, all the attorneys concurred with delegating final approval to the two outside specialist attorneys, Max H. and Rob W.

Speaking with them after the meeting, they told me that if I could satisfy one concern of Max's, they would approve our program. The entire business and all the investment we had made in time and money was on the line that weekend. After two days of grueling grilling by 19 attorneys, I impatiently asked, "What concern?" Max said if I could show him that one vacation booking point could be tied to a fraction of a square foot of a physical timeshare unit, he would approve it no matter how infinitesimal the part of a square foot might be. This

The Role of Simulations

I again wrote simulation programs to predict the likelihood of members' success at getting their first choice when using their points. There are at least two ways to simulate a process.

One approach is to use a looping program in a language like Fortran. You can create a series of if-then statements based on randomly chosen probabilities, test each action's likelihood, track the occurrences of each branch, and repeat the loop many times. Here is an example, if you are interested:

Consider simulating a coin flip. You would develop a computer program that generates a random number between 0 and 1. Based on this number, you can determine whether the coin lands on heads (<.5) or tails (>.5). Here's a simplified explanation of the simulation steps:

1. Initialize a counter for tracking the number of coin flips.

2. Pick the number of flips to simulate (e.g., 1000).

3. For each flip, generate a random number between 0 and 1.

4. If the random number is less than 0.5, count it as a heads flip. Otherwise, count it as tails.

5. After all flips have been simulated, report the count of heads and tails.

This simulation would provide insight into the expected outcomes of flipping a coin 1000 times. It will not always be 50:50, because actual results can vary due to the inherent randomness of coin flips. Nonetheless, simulations like this can offer a useful understanding of the probabilities involved, especially in more complex processes.

In simulations, random numbers can be uniformly distributed (like 0 or 1) or follow a distribution curve, such as an exponential curve, where certain numbers are more likely to be generated randomly. Simulations can replicate most processes. A "do it yourself" simulation involves creating a transaction, running it through its process, and observing the various outcomes that occur with different frequencies.

The second approach is to use simulation software. Years ago, I discovered a program called IGrafx Process, developed by the Texas company, Micrografx, which provided an alternative method for charting and simulating processes. Created in the early 1990s, I quickly embraced and adopted this software. It is a marvelous tool once you get familiar with it. However, it does require familiarity with flowcharts, and I recognize that not everyone relates to flowcharts as readily as I do. If you are interested, the key features of the iGrafx Process program include:

- Process Modeling: Users can create detailed models using standard process notation, including flowcharting symbols.

- Simulation: Users can simulate process behavior and test "what-if" scenarios to identify potential issues and optimize processes.

- Analysis and Optimization: It has analysis tools, such as process discovery, value stream mapping, and process performance analytics, to help users identify inefficiencies and areas for improvement.

- Reporting and Dashboards: It has reporting tools and customizable dashboards to help users monitor performance and identify trends.

• Collaboration and Governance: It supports collaboration and process governance, including version control, workflow management, and audit trails.

I've only needed and used the first two features above, which are powerful by themselves. At Stat Tab, I frequently utilized flowcharting to optimize processes for companies and departments and design computer programs. That experience was invaluable for learning this tool and simulating transactions. This program allows you to flowchart a process and then run transactions through the chart to see the outcomes. While running, it can even show you where the transaction is in the chart.

Using this program, I developed a unique technique for optimizing business processes. I always started my charting from an initial customer contact point. I then followed the steps taken within the company's internal process to handle the customer's request, ultimately ending with a response back to that customer. This end-to-end approach, repeated for every type of customer request, helped me pinpoint potential areas of breakdown and build a reporting system around critical points to alert management when intervention was necessary. In flowchart language, I learned that wherever there was a decision symbol or a process box that would take time to complete, a reporting system was needed to track frequency, performance, and backlogs. Overall, this approach is potent and can optimize any business operation.

I utilized this tool throughout my career to optimize and permanently document processes, such as the Collections Process. This documentation could then be used as a reference for similar processes. One could build an entire consulting business around using this tool to reengineer enterprises - a path I pursued.

My initial application of this tool was to simulate and optimize customer service processes for DVC. Subsequently, I applied it to various other business processes. While the recipients of the results might have yet to understand the tool's mechanics fully, they appreciated the clear conclusions it produced. Despite its initial learning curve, I found it an exceptional resource for reengineering business processes.

Mind Mapping

Speaking of charting, mapping, and capturing processes, I solved most of my Dyslexia problems when I read a book called "The Mind Map Book" by British author and consultant Tony Buzan, who popularized the concept of mind mapping in the 1960s. The idea came from Buzan studying geniuses and how they doodled while learning.

Mind mapping is a visual technique used to organize and structure ideas in a diagram called a Mind Map, which starts with a central idea and branches outward to related concepts in a hierarchical or radial pattern. It is used for note-taking, brainstorming, problem-solving, outlining, and planning, among other things, and can be done on paper or with specialized software tools.

Buzan's book describes the concept of mind mapping as a powerful tool for organizing and structuring ideas, notes, and information to make it easier to understand, organize, remember, and communicate. Since its creation, Mind

Mapping has been used in education, business, and personal development. Mind Maps can be found that already document many business processes. Between IGrafx and Mind Mapping tools, any process can be recorded, organized, optimized, and even perfected. And both flowcharts and Mind Maps can be used to train AI systems if they are not already.

I used these tools to document many processes during my career. These two tools enabled me to create the soundest and most thorough processes and ensure top-level customer service for DVC and others.

Member Reservations

We made a deal with the Disney World Park IT management to use their Reservations Systems so that we would be integrated with all their operations. This agreement entailed a complex negotiation because the Park people did not like the idea of Disney timeshare, and integrating would require many functional changes and integration.

We did accomplish those changes over the next 18 months. As done at RCI to mitigate risk, we investigated existing systems we might acquire, including the RCC system considered while I was at RCI. I settled on a Marketing and Sales system that a single individual owned. This system would also be able to print contracts, keep track of inventory available for sale and sold, and many other timeshare-specific jobs. This system had to be integrated with the others.

He did not want to sell the system, nor did we wish to license it. We finally settled on an exclusive license among timeshare resorts that sell or might sell a points-based product. DVC retained a right to deny any sales to a competitor, provided we pay him a fee for the "turn-down." The programming worked for our needs. Later, after I left DVC, those that followed me did not honor the contract, and a lawsuit followed – twice before the system owner finally got DVC to keep the original commitment. The original agreement was fair.

DVC Selects an Exchange Company

For any company selling timeshares, unless they already have a network of twenty or thirty resorts covering all the destinations where people like to vacation, the inclusion of the services of an Exchange Company in the sale is essential for overcoming the most common objection, which was, "But, what if we do not want to come back here every year."

The Disney Vacation Club Exchange contract was the gem of the industry in the eyes of the two Exchange Companies and therefore became a significant objective for RCI and their competitor Interval International (II). Undisputedly, Disney was the most prestigious account in the industry.

Because of my prior experience at RCI, I was assigned to run the bidding process and drive our Exchange Company selection. I was trusted to be objective, which I always was determined to be. Whichever company got our business, the Disney account would significantly boost its reputation and help them sign-up other timeshare developers. I expected both companies to fall all over themselves to get our business.

I prepared a formal bidding and negotiating process by laying out our requirements. I enlisted a Disney Attorney to escort me through the process, so I could not be accused of favoring or disfavoring either company. I was barely familiar with II, so I would mainly compare their services to those of RCI, which I knew very well. The main difference, as I would learn, was that their matching system, which assigned one of three colors to a deposited week, was far less sophisticated than the RCI internal point system I had designed. Swapping a DVC week for a week at another resort through II might be less than equal trade than through RCI. Also, RCI had twice as many affiliated resorts as II.

There were also fees to be negotiated, as we planned to include the Exchange Companies' annual membership fee in our annual club dues. Our one-stop-shopping goal was one fee and one call to do it all.

When it came time to meet with RCI, who sat across the table from my attorney and me? It was my original boss, Christel DeHaan, and my last boss at RCI, Bill M. - who was only briefly my supervisor and the one who impatiently terminated me. As we discussed a potential affiliation, my attorney, whom I had prepped thoroughly, was masterful. At one point, Christel turned to Bill M. and gave him the dirtiest look, which I interpreted as, "How could you have gotten us into this mess?"

We ended up signing with RCI because their system was fairer and had a much larger system of affiliates. However, I gained respect for II, which would benefit them later in my career. The other consideration was that Exchange Companies only traded full weeks, which reduced the flexibility built into our point system to book by day.

Months later, after the RCI contract was signed, I negotiated an agreement for DVC Members to make direct exchanges between our resort and a specific non-affiliated resort, bypassing RCI. This arrangement was the first time anyone had ever done a direct exchange because the standard affiliation agreements from both RCI and II forbade it. I believed, as did others, that the clause in their affiliation agreements violated anti-trust laws, but no one ever challenged them on it. In our case, our contract allowed it.

When the word got around the timeshare industry and back to RCI that we had signed directly with a resort, I received an angry call from Christel telling me I was violating our contract and that she would sue me. I pulled out my copy of our agreement, referred her to the relevant paragraph, and read the section that permitted DVC to do direct exchanges. She slammed down the phone. I told you our attorney was brilliant!

DVC-ACD

Just as acquiring or replacing an ACD had been needed at both United Airlines and RCI, DVC also needed one. In this case, the options boiled down to two: The Aspect Product that I had favored from my prior experience, or a product offered by Vista-United Telecommunications. Vista-United Telecommunications was a telephone company 50% owned by Disney and 50% owned by United Telecommunications, an experienced telephone company. Because Disney owned half of Vista-United, I could not reasonably choose a competitor – unless their ACD was functionally superior. So, I again planned a formal Request for Proposal (RFP) bidding process, laying out the selection criteria and asking both companies to respond formally to our requirements. Vista-United did not take kindly to my putting the business out for bid because they felt it should be automatically theirs.

I cleared with my boss, Mark Pacala, what I planned to do, knowing full well that internal political pressure would be headed our way. After comparing the two ACDs, I found the Vista-United comparable to the older technology like the Rockwell-Collins United Airlines used before I replaced them. Additionally, they lacked the feature I developed with Aspect, which let callers leave their number and receive a callback, thus avoiding hold time. I considered this feature mandatory for whichever system I selected.

As I was closing in on my decision, my boss got a call from Disney Headquarters in California and asked me to hold off long enough to speak with a corporate VP flying to Orlando to discuss the matter.

The VP showed up in my office with the President of Vista-United. They asked me why I leaned toward Aspect, and I told them my concerns about their ACD. They tried browbeating me, but I did not fold. Finally, they asked me what they could do to get me to change my mind. I said, "It seems that the only way I would take yours would be if it were free." My mind was made up.

That should be the end of the discussion. It was, and the two executives walked away.

But in a few minutes, they returned and said they would give DVC the ACD for FREE. I was puzzled by that offer but felt it would make up for the loss in efficiency compared to the Aspect. DVC got a $350,000 ACD for nothing. My boss gave me a plaque for that negotiation.

However, I believe what happened "under the covers" was that Disney Corporate paid Vista-United from a budget in California, either all or part of the cost, so the transaction was just a paper shuffle. But it still looked good in DVC's books. And this was not the only contentious relationship between DVC and the rest of Disney.

DVC Billboard

About a month before we opened sales, we erected a giant billboard on the drive into Disney World. The next day, it was gone. It was an enormous mystery at first. We then learned that a theme park executive, whose name, if I recall correctly, was Vice President of Disney-MGM Studios Bob Wies, had ordered it to be cut down. It was found crumpled up behind a warehouse. The billboard had cost $250,000 and was now in a heap.

When our DVC President, Peter Rummell, complained to Michael Eisner about the billboard being taken down, Eisner ordered it back up. The factions within Disney that opposed the idea of timeshare becoming part of their brand were behind the billboard's removal. The incident demonstrated the company politics at play during the creation of the Disney Vacation Club (DVC). However, our DDC President emerged victorious in this battle. Ultimately, DVC was highly successful and earned its place as a valuable addition to the Disney portfolio, generating significant profits.

Rummell was an excellent developer, but I had one funny experience with him. He called a Brainstorming meeting to generate some new development ideas for Disney. Surely you remember the rules for brainstorming:

1. Encourage open and non-judgmental participation – no idea is a bad idea.
2. Focus on generating many ideas.
3. Welcome diverse perspectives.
4. Build on ideas shared by others.
5. Minimize distractions.
6. Set time constraints.
7. Capture all ideas.

When the meeting commenced, there was a gap of silence, so I took the initiative to share an idea. Since Orlando offers all types of vacations except snow skiing, why not construct an indoor snow skiing facility to attract that specific segment of travelers? Surprisingly, Peter immediately dismissed my idea and quickly redirected the discussion, seeking other ideas from the others. Of course, this concept was successfully implemented in Japan a few years later.

DVC Marketing

When DVC opened for sales, everything was in place technically, but, to my surprise and our president's, our sales operation had not set up a marketing operation. We had built a beautiful Sales Center, which included a large, illuminated map with sets of lights to highlight US and International exchange destinations and boards where our sales agents would explain the sales points and Member Rules. However, the process for finding potential buyers who would be lured with incentives like free park tickets to visit the sales center and see the condominium model units were missing.

In other words, no mechanism had been created to attract families to tour the resort besides DVC having a few "soft sell" kiosks in Disney Village. Our request to Park Management to place hotel lobby desks was declined, not surprisingly. The lack of marketing was painfully evident after the first day. Our president asked me to quickly assemble a telemarketing operation, which would call families arriving at Disney hotels and offer them a tour of Disney's latest project.

Since we had installed small cubicle desks for each salesperson in the back of the Sales Center, each with a phone, we had a ready-made place for an outbound telemarketing operation, provided we operated it only in the evenings and could staff it. We worked with Disney HR to offer part-time evening work to any Disney employee who wanted to make a little extra money. We also obtained the inbound arrival guest lists from the hotels. Within only two days, we had overcome a massive oversight that might have been fatal to the project. Surprisingly, a sous chef from one of the Disney restaurants was our best telemarketer. Over the next few weeks, we worked out the kinks for a permanent call center operation and a permanent staff.

While planning to hire the Sales Team, we received much pushback from Disney and their HR department about what kind of people we should hire. They feared we would employ the "typical high-pressure timeshare salespeople." Instead, they insisted we hire "clean-cut," young, "Disney" people. We did as they asked with a single exception: we hired one experienced timeshare sales guy as a test. After the first week of live sales experience, guess who the top salesperson was, by far. Of course, the experienced timeshare salesman – because he knew how to handle objections, "paint pictures" of the family enjoying their vacations, and how to close. After that, we hired and trained the best talent for the job. In the first year, DVC completed $75 million in sales, and they have grown yearly.

Near the opening of this project, DVC was starting work on two other locations, one in Hilton Head and one in Vero Beach. I knew enough to anticipate that both would struggle and told my boss so, but the development department of DDC was committed. I expected the Hilton Head resort to be too small to generate substantial sales volume. When the land was maxed-out, they would have a maximum of 123 units with no room to expand beyond that. The second, Vero Beach would end up with 175 units. As I recall, it had land for additional units, but they were never built. Vero Beach had no nearby attractions, theme parks, or hotels to draw potential customers to tour. Neither property had opportunities for OPCs (Off-Premises Contacts) close enough to draw people to the resort. Even telemarketing would not work well as no Disney Hotels were near the property. The only hope would be preselling travel packages, redeemable for the limited number of available condominiums yet to sell out.

After refocusing on Orlando, the Disney Vacation Club constructed 2,800 units within the theme park acreage. Disney Vacation Club has since expanded by building or acquiring properties in California, Hawaii, and South

Carolina. As of September 2021, Disney Vacation Club has 16 resorts with over 4,500 vacation ownership units, serving over 220,000 members. The company achieved impressive sales, selling over $7 billion in timeshare points as of 2021.

Competitive Reaction

As soon as DVC demonstrated the success of selling Points, the other major timeshare companies took notice. Points Programs became the primary topic of the next American Resort Development Association (ARDA) conference and all those that followed. The first top major timeshare competitor to adopt a Point System was Hilton Grand Vacations Club (HGVC). They hired the same timeshare specialty law firm that we used.

At the conference, a session was conducted about Point-based Timeshare products. The session consisted of a panel of "specialists" who spoke briefly about their expertise. We were not invited to be a panelist but would not have refused if they had, as we had no interest in helping our competition.

One of the panelists was one of the attorneys I had worked closely with to develop DVC's points program. I will refer to him here as "Bob." He had been advising HGVC on the creation of their club, and the Hilton Club was the subject of his comments on the panel.

After the speakers had finished, the Moderator opened the floor to questions from the attendees, at least 500 in number. When a couple of questions had already been asked and answered, I raised my hand to ask a question. I was standing below them to their stage left. The Moderator recognized me, so I asked, "I could not help but notice the similarities between the HGVC Points program and the DVC program. Can you explain how this came about?" The Moderator then turned to "Bob" and indicated that the question was for him. "Bob," turning toward me, responded by "flipping me the bird" in front of those hundreds of attendees. He followed with a vague and evasive response. I could not help but smile.

He was obfuscating that they had utilized work that DVC had paid them for and provided it to one or more of our competitors, potentially charging them for the creative work we had done. Although I found his response amusing, I still believe that the law firm exhibited inappropriate, unprofessional, and unethical behavior, along with improper business practices.

Other Notable Actions

Disney's brand name carries significant weight with vendors, especially regarding partner marketing, which I leveraged to reduce costs for the DVC condominiums. As I did with the Apple computers at Euro Disney and the ACD, I negotiated with General Electric, an existing client of EPCOT, to provide lightbulbs for the 250 units we were constructing in exchange for a placard in each unit promoting the use of GE lightbulbs. This business strategy, product placement, is a common practice between Disney and its sponsors to promote merchandise to millions of Disney visitors. However, GE had not calculated the number of lightbulbs they were committing to until we requested delivery. As it turned out, the required lightbulbs amounted to $85,000, which surprised and frustrated GE when they realized the extent of their commitment. Despite their initial objections, they fulfilled their promise. I was confident in the quality and value of our product and did my best to add value to DVC.

I also negotiated with the Theme Park Hotels division to allow DVC Members to redeem their points for bookings in the Grand Floridian, Disney's finest hotel. This arrangement was a significant coup, added value to the product, and added great "sizzle" to the DVC sales pitch. We also built four park passes annually for ten years into the purchase price of the most popular DVC timeshare points package. This extra benefit added genuine value again to the DVC product, so much so that it could not be continued after the first few years of sales.

I was the first DVC member. My customer number was seven rather than one because they used the first six customer numbers for testing. I bought the first Interest to show my confidence in the Point System I had developed. I made the decision also because, as a company associate, I was eligible for a 20% discount. I also did it because of my conviction that if anything went wrong with the Club, Disney would stand behind it. Years later, I sold our membership when we stopped using points, mainly because we lived close to Disney World in Orlando.

DVC FIRSTS

At DVC, we accomplished these firsts:

1. A new multi-billion-dollar business for Disney, from the start of design to full sales in 18 months.
2. The first timeshare Point System for a Global Brand.
3. The first one-stop-shop Member Services system that integrated the DVC point management system, Disney's Reservation System, Member Services member tracking system, and travel agency services. However, we faced difficulties finding experienced Member Services agents within our budget. Consequently, we revised our approach to DVC agents transferring calls to Disney Travel while remaining on the line as necessary, a strategy that has proven successful.
4. The first implementation of multiple monitors operating from a single desktop computer on Apple MACs.
5. Obtained, for free, DVC's first ACD system and $85,000 in GE light bulbs.
6. Created the DVC marketing department in 2 days.

Years after I left DVC, while still a Member, I went to the local Disney hotel where DVC held its annual DVC Members meeting. I was curious about how the Club was doing. After the meeting, I introduced myself to some current management, including the president. Surprisingly, they were familiar with my "Issues of Substance" document and "Max's Comfort Analysis." Both had become corporate legends. I thought it funny that I had become part of DVC's "permanent record," like my experience at United Airlines with Apollo. Well, semi-permanent record. They may have forgotten by now.

What I learned at DVC

At DVC, and having to take on responsibility for Marketing, I learned a clearer distinction of the difference between Marketing and Sales.

Marketing is the process of attracting prospects to engage with salespeople. It involves creating specific awareness, promoting the brand and its benefits, and providing calls-to-action to encourage prospects to engage, while not straying into the presentation that will be made by Sales. And Marketing is different from public relations (PR), which is focused on promoting the brand and creating general awareness. Marketing locates qualified prospects and gets them in front of salespeople.

While marketing is necessary in most cases to generate opportunities for sales, there are exceptions. For instance, highly sought-after products like the iPhone at its peak demand required only awareness to generate sales.

In general, salespeople are responsible for meeting, greeting, and understanding potential buyers' needs, linking them to the product benefits, painting pictures in the buyer's mind of them experiencing the product, and guiding them to a decision. Salespeople have different personalities and strive to form connections, build relationships, and discover the motivations of customers, all with the goal of guiding them to a decision. They say, "make a friend to make a sale."

A common marketing mistake is attempting to do the job of sales and failing because it is too early in the buying process. This happens because the potential buyer has not been brought along enough in the decision-making process.

These are common successful salespeople's characteristics:

1. Empathetic: An effective salesperson is empathetic -- they understand and relate to the customer's needs and feelings.

2. Confident: A successful salesperson exudes confidence, as it inspires trust and builds credibility with the customer.

3. Resilient: Sales is a tough job, and an effective salesperson needs to handle rejection and bounce back from setbacks.

4. Assertive: An effective salesperson is assertive in their approach, without being pushy or aggressive. They take charge of the conversation and guide the customer toward a purchasing decision.

5. Personable: Successful salespeople are personable and approachable, making customers feel at ease.

6. Inquisitive: An effective salesperson is inquisitive, asking thoughtful questions to understand the customer's needs and motivations.

7. Resourceful: An effective salesperson finds creative solutions to meet the customer's needs.

Effective salespeople aim to convert as many prospects as possible into customers, or in this case, members. Successful marketing specialists exhibit different personality traits compared to salespeople. Here are common effective marketing specialist characteristics:

1. Creative: Can think outside the box and develop innovative ideas to attract qualified prospects.

2. Analytical: Use data and insights to understand customer behavior and make informed decisions.

3. Adaptable: Be adaptable and willing to change strategies as needed, as Marketing is constantly evolving

4. Collaborative: Can work closely with sales and others to achieve common goals.

5. Strategic: Can develop and execute marketing plans that align with the overall business objectives.

6. Excellent communicator: Can convey opportunities to learn about the product.

7. Detail-oriented: Ensuring all marketing materials are high-quality and error-free and include a call-to-action (getting customers to do something right now). Effective marketing specialists, equipped with the above personality traits, create and manage successful campaigns that drive sales growth.

How different the skill sets are from one another.

The End Cometh Again

After all the systems were operational, the need for my creative and technical skills at DVC dwindled. Corporate Disney World IT did not require my expertise, and my only other offer was a lower-paying Project Manager role at a new DVC resort that I knew would struggle. Seeing no further opportunities to leverage my abilities, I decided to leave despite my boss's wishes.

Before resigning, I contacted Herb Alfree, the former VP of Marketing at RCI and a colleague. Herb had an extensive network, and I knew he would help me if he could.

CHAPTER 7
FOUR SEASONS USA

Lake Ozark, MO

When I contacted him, Herb was simultaneously a consultant to several developers, including a husband and wife who owned a large land tract in Lake Ozark, MO. They called their property Four Seasons. He was in the process of helping them with their development plans and with marketing and sales. They needed someone full-time onsite as COO, which Herb did not want to do. So, I took the job on a consulting contract and moved to Lake Ozark. My wife and I moved into a company-provided home on the lake. The driveway was steep and treacherous. So perilous that the garage with a retractable door on both ends because backing up was dangerous, anytime, and impossible in the winter. Missouri is known for ice storms in the winter as well as snow.

Because Four Seasons used the same name as the well-known luxury hotel chain of the same name, we had a potential conflict in branding. The property owners in Lake Ozark had used the name Four Seasons before the hotel chain but could only claim the right to use it within Missouri. So, as we did not want customers to think the property was of a luxury level, even though it was decent, and we wanted to distinguish ourselves from the hotel chain, we changed the name to Four Seasons USA.

The wife had inherited 7,000 acres from her father. Her husband ran the business aspects. The property is wrapped around a large section of Lake Ozark, on both sides of the lake, which was formed by a power company damming up a river many years before. The original owner (father) paid about $0.50 per acre when he bought it years before (just before the dam went in).

Over the years, the family had built a hotel, the Lodge of the Four Seasons, golf courses, and an extensive indoor and outdoor tennis club. They had sold many single-family lots, some condominiums, and a dozen or more "tree houses" near their Hotel. The property was about 3.5 hours from St. Louis and Kansas City. This location was the perfect drive-to vacation destination for those two large metropolises.

A Marketing and Sales operation was already in place, run by a consultant named Russ Brown, but financing was hard to arrange, as it was the era of the late 1980s Savings and Loan Crisis. So, getting construction loans was impossible and, if even accomplished, would be very expensive. And the company was struggling financially. Our mission was to "sell our way out of debt and into prosperity."

When I arrived, a "controller" was already on staff, reporting to me. He was messy, disorganized, and over his head. He supervised the accounting department, which was in a shambles. I immediately hired an experienced CPA to take over the accounting department, reporting to me. He quickly identified that our company could no longer order on credit with many vendors and was about $450,000 in arrears. With Herb, we pitched the owners on providing us with a cash injection. They had some available cash but only slightly more than we were asking.

They reluctantly agreed. I went back to Florida on Thursday for the weekend to bring my wife up. On Monday, I learned that the owners had delivered the check on Friday after I left and that the "Controller," who did not handle stress well, had immediately used it to pay down all the past-due vendors, so they would stop stressing him. I was furious because we could have operated off those funds for months with careful cash flow management. As it was, we were right back in the hole we had been in, with no available cash, but now with vendors who still cut us off for fear of us becoming delinquent again. I would have paid them some each with the understanding that they would continue to work with us. My opportunity to negotiate some breathing room was gone.

In the eighteen months we worked there, we successfully designed, implemented, and sold four new projects in the Ozarks and an additional one in Branson. We restarted their timeshare sales operation in Lake Ozark at one property and started a second development near the Hotel. We rejuvenated their sales of lots, condominiums, and single-family homes. We built a whole community of lots across the lake and a separate fractional project next to the clubhouse.

We were the first to utilize an "Asset Light" development model. Later, many developers would adopt this model. When Herb and I created our model out of the company's desperation, we were unaware of anyone else using a similar model. Ours was the most innovative approach.

Here is how Asset Light development worked for us: We arranged with a local builder to build fourplexes that we would sell as timeshare and pay him a portion of each week as it was sold. He would build one unit, and when we sold 75% of the weeks in that one, he would start building the second, and so on. The builder's share was easily auditable and made his profits much higher than he would make building and selling whole-owner condominiums, especially when properties were barely selling. He did not know anything about timeshares, but we did. We had a construction department but no funds to build units ourselves. This arrangement was a great example of a symbiotic relationship – two businesses succeeding in an impossible banking environment.

We made a similar arrangement with his wife, who owned a furniture business, to furnish the units. Using this Asset Light approach, we kept our company and all the local supporting businesses going during difficult economic times. Furthermore, this approach enabled us to avoid dealing with banks or having to improve the company's poor credit reputation in the community.

First Fractional Development

We also developed one of the first successful fractional resorts in the country. We designed two-story homes with spectacular views overlooking Lake Ozark. We built them along a ridge next to the tennis club. Again, we used the Asset-Light approach. We sold each building as our partner constructed it, starting construction on the next building when three-quarters of the prior building was sold. Each home would be owned by thirteen owners, each owning four weeks, one in each season. I developed a calendar that rotated the specific weeks to ensure equal access to every week of the year for all owners over time. Although we could have structured the interests differently, such as fixing the weeks to be the same every year, it would have meant different prices for different interests and the possibility of certain interests selling more slowly, so we opted to keep the pricing and therefore the demand consistent for everyone by rotating the weeks.

The owner became an RCI Member and could exchange unused weeks for a different destination through the company's pre-existing Exchange Company relationship with RCI.

Our Fractional product was immensely successful due to several factors. Firstly, the property was located within a few hours' drive of two major metropolitan cities. The construction quality was top-notch, and the property boasted stunning views, which we maximized with floor-to-ceiling living room windows.

In contrast, other Fractional projects attempted by different developers in fly-to destinations failed to gain traction. Potential buyers were not interested in purchasing a resort where they had to fly to the destination four times a year.

Land Development – the Lots Across the Lake

Half of the company's land for development was on the other side of the lake. However, this location was off the real estate market because it was inconvenient to access - potential buyers would have to drive 30 miles each way to see it.

I devised a solution to overcome this limitation and create demand by buying at auction an auto ferry and hiring a Mississippi-River-licensed captain to operate it. Our construction department built concrete ramps on both sides of the lake, allowing the ferry to transport customers and their cars to and from the other property in just a few minutes each way.

Next, on our terms, I sought out an infrastructure company based in Kansas City willing to install the necessary infrastructure for land development on the other side of the lake. This deal included roads, curbs, sidewalks, sewer, electrical, telephone, and streetlights, everything needed to create buildable lots.

For this project, I applied a variation of our Asset Light model. The infrastructure company agreed to be compensated out of the proceeds of each lot sale in a manner that would enable them to recover their investment, with a generous profit, by the time the first half of the lots were sold.

All the Asset-lite deals I made depended heavily on trust between the parties. In this case, my trust was linked to the son of the owner of the Kansas City company, with whom I made the deal, and my gut was telling me that it might be possible that his father might move in "shadowy circles" known to exist in that city. I did not want to test that theory, so I made doubly sure payments were made on time.

By implementing this strategy, we could generate substantial cash flow and profits, mainly since the owners had acquired the land for just $0.50 per acre. Furthermore, the possibility of constructing a new bridge to the area was crucial in persuading hesitant buyers to purchase. The bridge was built a few years later, so the early buyers received an excellent bargain due to the newfound accessibility and convenience.

Golf Course Home Lots

The company owners requested that our Marketing and Sales team focus on selling more lots alongside the golf course. Unfortunately, these lots were significantly less attractive than those with lake views, and we encountered significant difficulty selling more than just a handful. Moreover, the area looked barren and uninviting due to the limited number of homes built along the golf course. Potential buyers often prefer to have neighbors nearby, but most of these lots had none. Additionally, the lack of homes on the golf course raised concerns among potential buyers who wondered if there was a problem with the development.

Our Sales VP approached me with a request to offer a discount on lots for a significant weekend sales event. As their manager, I recognized the importance of listening to the sales team and providing them with reasons to feel motivated and energized. I knew that such a discount event would excite the sales team, so I agreed to the proposal on the condition that it would be a one-time sale and that buyers would commit in writing to begin construction within twelve months. Our company needed the cash flow to meet payroll, and I believed the discounted sale event would generate more revenue than a typical weekend.

We had a successful weekend with a significant boost in sales. However, the company owners were dissatisfied with my decision to discount their "premium" golf-course inventory. They had an inflated sense of their lots' values based on comparing the property to thriving golf communities in other parts of the country. However, they should have considered that these other communities were situated in residential areas, close to where people worked, rather than drive-to resort areas like ours. These lots also faced intense competition from the highly popular lakefront lots, which were the focal point of most Ozark vacations. As a result, the market value of the golf-course lots was less than what the owners believed, and they never fully appreciated that the discount strategy was necessary to generate the cash flow we badly needed. We had taken a risk, and it had paid off to keep the company afloat.

Marketing and Sales Computer

The Four Seasons USA reservations system ran on a small AS400 with Unix. We added a JD Edwards accounting system. They also had a primitive telemarketing computer system for selling weekend packages to get people into the sales pipeline. I hired Chuck Priller, one of my prior programming managers who had left RCI. He had recently been divorced. He quickly straightened out the system and the telemarketing operation. Chuck showed me a bit about Unix, which I had yet to encounter. Later that year, I became concerned when I learned that Chuck was dating the local mayor's wife. I sensed trouble. However, my concern was alleviated when I heard that the mayor had no problem with it.

Another problem I dealt with was the complete absence of any printed sales brochures or marketing materials. The company had no library of photo images, and every printed piece was woefully outdated. There were no posters, flyers, mailers, or collateral.

To address this issue, I enlisted the help of Jeffrey Bruce, my designer friend who had assisted me in Scottsdale. He and a professional photographer came for a week and, using a helicopter for a day, captured the finest photo shoot of the resort that I had ever seen or would. The resulting photos were high-resolution and expertly composed. The portfolio

Jeffery delivered allowed us to create top-quality collateral for marketing and sales finally.

Although the property owners were initially unhappy with my decision to spend money on photography, they eventually admitted that the new collateral significantly improved the resort's image. Everything with this business was risky and challenging.

The company owned undeveloped land on a key intersection, and we had the opportunity to construct a new welcome center if we could get financing. A welcome center on that corner would be a huge draw for customers. The intersection was the corner of the road leading to our resorts and the main highway that Ozark visitors used to enter the town of Lake of the Ozarks. Most of our resorts were in low-vehicle traffic areas and mainly visited by those who already owned homes on the property or were hotel guests. This new center would act as a draw for all vacationers entering Lake of the Ozarks and offer an ideal chance for us to attract many more visitors to view our resort properties. But we had a challenge.

Our challenge was obtaining financing to build the welcome center, as banks were not lending money then. However, a local contact who knew the bankers and I suggested the bank might loan us the necessary funds if I were open with them. Despite the risk of telling too much and being denied, I decided to follow his advice.

When I met with the bankers and showed them our financials, the bank president told me they already knew how bad things were at our company and appreciated that I was willing, to be honest and forthright, and they would trust me. The company seemed to have been in denial about its condition until then, so the bank had previously refused to loan them any money. They granted my request for the loan, and we built the welcome center. It was not finished until after I left but it was a huge success.

A Missouri Near Death Experience

Missouri is a scary place. Right-wing extremists and pirate radio stations are blaring out instructions on attacking Federal troops. We heard in one transmission: "… pick a cloudy day, so there is no overwatch, sneak up behind a soldier, hit him in the head with a metal bar, and take his gun and ammo." Not joking; we heard such a broadcast. If you would like to get a sense of what Missouri is like, watch the Jennifer Lawrence movie *Winters Bone*. It was filmed between Lake Ozark and Branson, and I think it will give you a feel for the area. That movie certainly conveys how I feel about Lake Ozark, MO. I had already been given one warning that something terrible might happen if I did not give the construction crew the day off for the first day of deer hunting season.

My wife and I had a car we did not need, so I put a sign on the company bulletin board offering the vehicle for sale for $2,000. That was an excellent price for the car, considering its condition. My wife wanted it to go to a deserving family as it was in excellent condition. A man who worked in maintenance for the hotel, whom I did not know, called, and said he wanted the car. A couple came and paid $500 from the wife's mother in cash, and the man provided a check for $1,500, which bounced. When I contacted him about the payment, he told me to go to hell and that he would kill me if I called him again. After I hung up, I reflected on the situation for about 10 seconds before deciding that the money was not worth the risk of the guy being serious. When I checked on his employment, he had already left the company. We heard of people disappearing, so why take the risk?

Early Branson

Herb and I would fly to Branson by small plane one day per week to work simultaneously on starting a new timeshare resort in Branson. Herb and I worked as consultants, not employees, so this project was for another financial backer, not the Four Seasons USA owners. The potential demand for timeshares and the vacationer traffic were higher in Branson than in Lake Ozark, and the competition for timeshares in that area was still low. At the time, there was only one other older timeshare resort in Branson, Missouri. Land speculation was rampant, with one parcel of land I knew flipping owners three times in one week.

I remember standing on a hillside with a man whose name I have forgotten, looking out over what was then all of Branson, and seeing new construction as far as the eye could see, across the entire area in front of me. The man asked me whether I thought all the construction going on before our eyes would be a major disaster or a huge success. I replied, "With all the people investing so much money in all this construction, they cannot; they will not let it fail. So yes, I think it will be a huge success."

I remember discussing Branson's potential with Herb on our flights there and back. The flight to the south, which covered about 100 miles and departed at 6:30 AM, was always heavily turbulent and took an hour. In contrast, the flight back around 5:30 PM was always smooth and took about 45 minutes due to prevailing winds.

The Branson property had an existing clubhouse that would serve well as a Sales Center. We built condominiums at the bottom of the sloping land behind the clubhouse directly on Lake Taneycomo, a dammed river created by the Table Rock Lake dam outflow. The units were the same fourplexes we built in Lake Ozark, using the same builder and Asset Light model.

As I mentioned above, Branson was a hotbed of land speculation at the time, with hucksters flying in every week to flip this property or that. The land we were to build on was threatened to be sold out from under us for months, but it never did because our use was the highest and best use for that land. During the risk of upheaval, we marshaled on, creating a sales and marketing team and leasing marketing locations (OPCs). I hired an accountant to process the money. The project was successful and is still operating today.

What I Learned

Once we had managed Four Seasons USA through a terrible business cycle and got them on level ground financially, they decided they did not need us anymore. I was okay with that because my creative juices were running dry again. But at least I had my first introduction to Unix and set up another telemarketing operation.

I had learned how to arrange creative financing when loans were impossible, how to arrange traditional loans by being forthcoming with bankers, and how to be a developer, running everything from the construction department to marketing and sales.

So, my wife and I left and, on the drive east, debated whether to head to Chicago, our hometown, or go back to Florida, where we still had our house. Florida won out, even though the job prospects were unknown.

Firsts at Four Seasons USA

We achieved several Firsts.

1. Stabilized their computer systems, including adding support for Branson.
2. Built the financial models for three profitable Asset Light projects and managed them successfully.
3. Built the first Fractional Project and started a new timeshare project.
4. We opened lot sales across the lake by launching a ferry service.
5. We built and sold the only new timeshare resort in Branson, MO, at that time, and the second ever.
6. We created a small telemarketing operation.
7. Planned, financed, and built a new Welcome Center
8. Learned all the essential aspects of being a property developer.

CHAPTER 8
MARRIOTT VACATION CLUB INTERNATIONAL (MVCI)

First MVCI Stint: Product Development

While working at RCI, I met a sales director for a Rhode Island developer named Pete Watzka. He impressed me on our first meeting, and he had since moved to Lakeland, FL, to work for MVCI, which was called Marriott Ownership Resorts Inc. (MORI) at the time. Since I had met Pete, Marriott had entered the Timeshare Industry by buying American Resorts Corporation, an existing timeshare company operating in Hilton Head, SC, which was co-owned by Bob Miller.

Bob Miller came over in the deal as President of MORI and later MVCI, and he brought in Pete as his SVP of Marketing and Sales. Next, Pete hired me as VP of Product Development at their offices in Lakeland, FL. It was a long commute from Orlando, but worth it to be among the early employees of this company which was on the verge of a massive expansion to new locations, with more extensive properties than most timeshare developers, with better quality, and Marriott's quality service levels.

My assignment as VP of Product Development was to build a team to produce a Product Development Strategy -- a Roadmap for future products. When the company moved its executive offices to Orlando, I hired a small team, including Rob Hebeler, Oran Greene, Ray Katkish, and Bob Phillips (from DVC). We would add or subtract others over the months but kept this core team.

As its first assignment, my group explored all the varied product possibilities for Marriott Vacation Club, including points, fractional, fixed and floating date weeks, and more.

One set of tools we developed was a set of standards, in the form of a pair of checklists, for creating new products. The standards covered all the aspects of a new product that a customer would want in a product. For example, at one point, I was assigned to a task force to create a Member Rewards Card, a fancy Black Card, to convey an elite status for the best members. I applied the new product standards to the initial design of the card, and the results indicated that the Black Card would not be very appealing to the company's members. Evidence to the contrary, the task force leader chose to roll out the card despite my warnings, and the card program failed as a new product.

Ritz Carlton Sales

At one point, MVCI was asked to provide a speaker for a gathering of 200 of Ritz Carlton's salespeople. Ritz Carlton was a division of Marriott Hotels, just as MVCI was. They wanted to know what their timeshare division cousins were up to. I felt it was essential to put timeshare in a good light internally as timeshare was still a dirty word to the public. As no one likes public speaking, the task was passed down to me, as happened when I was with Stat Tab. Since Ritz Carlton is such a top-quality brand, I wanted them to think highly of MVCI. Because I had given a talk to salespeople at RCI on quality, I felt a similar speech would be appropriate for Ritz Carlton and allow me to reuse some of my content from my RCI talk. For the third major speech in my career, I felt prepared. I did the Godiva gimmick again, not the Pig or the Reams of Paper bits. It went well, and again; I got a standing ovation.

The End Cometh Once More

After completing the Product Development masterpiece and presenting it to Pete Watzka, our SVP of Marketing and Sales, we realized that our comprehensive report and its contents were more than the company could absorb at its stage of development. The company was going through a consolidating phase, and our plan called for expansion. As our project was completed, the economy was turning sour, and with the company reorganizing and moving into a different office building, my team was furloughed. It made sense for the times, so the change did not bother me. I had confidence, and because I have always put at least 15% of my pay into savings, I had the wherewithal to withstand the lack of employment, at least for a time.

CHAPTER 9

ORANGE LAKE

In a matter of weeks, I found a new position as SVP with Wilson Resort Management Co at Orange Lake Country Club and Resorts, a local Orlando timeshare resort founded by Kemmons Wilson, the creator of Holiday Inns. It was a large resort, continuously expanding over the years. My responsibilities included IT, Sales Operations, Mortgage Processing, Timeshare Resales, Portfolio Management, Billing and Collections, Communications Systems, Attraction Ticket Sales, Accounting, and more. I was responsible for 120 people and a $10 million annual expense budget.

My immediate boss was an erratic leader, praising one minute, yelling the next, and brooding the next. I'm not a psychologist, but he seemed manic-depressive to me. I heard he once took a computer from his desk that was irritating him and threw it through a second-story window, nearly hitting an employee walking past. But, on the other hand, he was smart, could spot an error in a complex spreadsheet in seconds, and succeeded at growing the company.

Year 2000 (aka Y2K)

The company was concerned about the Y2K problem, which posed a threat related to how many computer systems and mechanical equipment stored years as two digits. Computer systems could not make accurate date comparisons when the year changed from 1999 to 2000 since "00" is less than "99," even though 2000 is greater than 1999.

This one-time logic flaw could result in numerous system failures. The company had several systems in place, including front-desk and reservation systems, satellite communications systems, elevators, and fire-monitoring equipment, so it was crucial to identify potential failure points. As a result, we had to research all the devices that might contain electronic chips or run computer programs to ensure they would function correctly during the transition to the new millennium.

With sufficient warning, I believed we would have enough time to enhance some systems during the process rather than merely patching them.

Upon arrival, I visited the programmers' area and discovered that they worked in a basement section, occupying cubicles. Adjacent to their workspace was the manager's office near a compact computer room that housed an IBM AS400. The team had four programmers and a manager who programmed, all in RPG III, a simple language but powerful enough and sufficient for their needs.

When I entered the computer room, I was shocked to see cables running haphazardly everywhere, even on the floor, with trash visible between them. It was an accident waiting to happen. I immediately asked the department manager to clean up the room, which he acknowledged was in poor condition. He worked throughout the weekend and proudly presented his work to me on Monday. He had done an excellent job reorganizing the cables into bundles and labeling them. His improvements prevented a potential calamity.

Given my experience and the president's distance, I was granted considerable latitude, something I appreciated. The resort was about to expand by another 1,200 units, doubling in size. We had to prepare the "ground" for this expansion. We had to install data and voice cables in the same trenches as electric lines that we were being put into the new acreage. The resort ran its own Cable network, with large satellite dishes, and the cables for the televisions in the units needed to be run and connected.

The smaller departments also had needs. The billing and collections department needed computer support – it was a small outbound call center calling delinquent members across the US in different time zones. Some laws vary by state regarding how late a bill collector can call, so the call lists need sorting by state and sometimes by city.

Oracle

The collections department served the Mortgage processing department. Because the timeshare business is a negative cash flow business, all the profits are made at the end of the mortgage payoffs. This financial structure means a product is built and sold this year, but the payments stream over three to seven years, depending on the sale amount and the resort. So, what timeshare developers do is borrow money from a bank against that payment stream, so they have money to build more units. These were traditional loans.

Some developers use a more advanced technique of "securitizing" bundles of mortgages. Securitizing means that a batch of mortgages was put into a trust, and shares in the trust were sold to investment companies.

Both types of financing, loans or secured trusts, required that when a mortgagee stopped paying on their note, we were required to remove it from the trust or loan and replace it with another one making payments. This way, the bank or trust owners were guaranteed the promised initial return on their investment. These processes required computer tracking, and this area was weak at Orange Lake.

So, I hired a sharp programmer, Jean-Marc Denis, who had worked for Arianespace, the French government's rocket company in South America. He had moved to Florida, hoping to get on at NASA, but they were not hiring then. He programmed quickly, which I appreciated. I gave him the assignment of installing the Oracle database system and to do so within two weeks.

Anyone familiar with Oracle would have laughed at me for that request. But Jean-Marc not only didn't laugh, he accepted the challenge. As he set up Oracle, I laid out what the mortgage tracking database should look like and coded the new reports we would need to manage the bank loan and securitized portfolios properly. The reports included aging and tracking delinquencies, substitutions, and defaults. With our team of two, we knocked out a sophisticated Oracle portfolio management system in two weeks.

Resales was a department of one -- a very sharp timeshare sales lady. Any member who no longer wanted their timeshare and contacted us to buy it back had to talk to Marie.

She would resell the members 40% of the time on keeping what they owned. Mortgages that went into default also went to Marie. Defaulted mortgages had to go through a "bankruptcy" process, where the timeshare interests were auctioned off on the local courthouse steps. Of course, the only bidder was Orange Lake, who repurchased them at a fraction of their original cost to resell them through a separate sales team.

All the sales (initial and resales) required computer and process engineering support, which we cleaned up through the Y2K project. In total, we reprogrammed over 1,200 programs. For extra certainty, we went through all the programs twice and tested them all twice. Our entire team was onsite the night the calendar rolled over to Jan. 1, 2000, in case we missed anything. We had only one problem -- a minor problem with a piece of Air Conditioning equipment that we did not know had date logic. It was dealt with that same night.

Around this time, I became friends with Ben, the son of my Administrative Assistant, Lois Silverman, when she invited us to see him in a local high school production. His performance was excellent. We had a mutual interest in movie production and joined the Florida Motion Picture and Television Association. I also became a columnist in a local entertainment newspaper, Focus-in, and dabbled in Screenwriting. This friendship has continued for over 16 years and is ongoing.

Colt Terry

Following a turnover, I hired an Administrative Assistant who embodied the qualities of a gentle Southern lady - polite, kind, and efficient.

One day on my return from lunch, I noticed that she was eating at her desk and reading a thick manual, which read "Special Forces Operations Manual." I immediately commented that I thought that was a very unusual book for her to be reading. She then told me that her husband, who retired from the Army, was one of the Original Green Berets and helped create what became known as Special Forces.

I was flabbergasted by such a contrast. She suggested that I meet him. His name was Colt Terry. She explained that he had a lot of fascinating stories to tell. She introduced him to me that afternoon when he came to pick her up. She was right about his stories.

My wife and I got to know them, and at one point, I suggested that he put his stories down in writing because they were nothing short of amazing. Unfortunately, he did not feel he could write them down, so I volunteered to try to write his stories into a book for him. We agreed on a 50:50 split of any proceeds.

Over the course of the following year, he and I collaborated. I interviewed him on tape and turned the interviews into a book titled Colt Terry, Green Beret. He assisted me in finding our first publisher, The Naval Institute Press.

Initially, since I had been writing articles in my spare time for a local entertainment newspaper called Focus-In, my approach to capturing Colt's stories was to write them as a series of newspaper articles. The publisher rejected that approach and asked me to rewrite the stories as a Biography, which meant writing it chronologically and including Colt's early life and life after his service days. It meant much rework. Then, after rewriting it, they had two military historians review it for fact-checking, which I responded to with Colt's input. Then 9/11 happened, and they withdrew their commitment to print the book due to the uncertainty.

After several months of searching and receiving 22 rejections, I eventually found another publisher - Texas A&M University Press - which was our last hope.

The pre-publishing work was grueling, making me cut out 30,000 words to tighten the writing and then later another 12,000 words to make room for photos. The publisher made it even more difficult by having it reviewed and challenged by two more military historians.

Despite facing four reviews by historians and numerous challenges, Colt's facts were proven to be accurate in every case except one. I had to research and find proof for every point they challenged. When I could not find definite evidence for that one, I removed the contentious reference from the book.

During my research, I also contacted Colt's XO and some of his Army contacts. One kept whispering that something in the book was not true. I finally had to corner Colt and ask what he had not been truthful about. After some time, he finally confessed that he had given a false description of how he was wounded in the foot on an island off North Korea. He had said he had been struck by a stray bullet from the mainland, which was not far away.

He finally confessed to what had indeed occurred. He had accidentally shot himself in the foot while demonstrating his fast draw to a South Korean cohort. Contrary to his previous claim, he had concealed the truth out of shame, fearing being labeled a malingerer. This notion was preposterous, as I regarded Colt as the bravest man I had ever encountered.

Very soon after the book, *Colt Terry, Green Beret*, was published, Colt underwent Prostate Cancer surgery, probably caused by Agent Orange. Colt remained optimistic and even sold copies of the book to his two surgeons before his surgery.

Sadly, he did not survive the operation. However, that writing experience helped me to learn the discipline of writing books and understand the publishing business.

What I learned

After reengineering all their critical processes, my tenure was up at Orange Lake. My several predecessors, I was told, only lasted twelve to eighteen months. Nevertheless, I managed to remain for three very productive years.

Orange Lake offers one of the best timeshare experiences in the industry aside from Disney, Hilton, and Marriott. It provides a great experience where a family can have a fun vacation without spending a lot. Many activities were targeted to modest budgets, such as miniature golf, reasonably priced regular golf, indoor games for rainy days, pool areas with expansive decks for sunbathing, and more made its members' vacations fun at a moderate cost. The ambiance reflected and extended the vision of Kemmons Wilson, the founder of Holiday Inn and owner of Orange Lake Resort.

I had the pleasure of meeting Kemmons when he visited the resort from his headquarters in Memphis, TN. The management team had been assembled for a Christmas gathering. Our president, whom I will refer to as Charlie, took the stage at the podium and introduced a few newly promoted vice presidents to the audience. He introduced Kemmons, who caught everyone off guard when he said, "One thing I like about Charlie is that he gives away titles instead of money."

I later discovered from Kemmons' autobiography that he always retained for himself all decisions about purchasing carpeting across all his companies because it was his business when he started his empire. He had bought millions of yards of carpeting and knew everyone in the industry.

I also learned more about Oracle software at Orange Lake than before. I had run into the president of Oracle years before at a conference where he was promoting the value of his relational databases compared to hierarchical databases like the Focus system used. I knew from my early programming days that relational databases may process faster while hierarchical databases may use less storage because there is less data duplication, depending on the specific application. I have used both and find that, with today's computers' power, it rarely makes any difference. Also, humans relate better to relational databases (pun intended) because they can appear like large Excel Spreadsheets. If you are interested in more details read below:

A relational database is a type of database that organizes data into tables or "relations" with predefined relationships between them. For example, a simple relational database for a school might have a "students" table with columns for student names, IDs, and classes and a "classes" table with columns for class names and instructors. A common "class" column would relate these two tables. When a new student is added to the "students" table or a new class is added to the "classes" table, the database management system automatically maintains the relationship between the two tables, making retrieving information about students and their classes easy.

Oracle and other relational databases provide for "joins," which means the computer software links a column of data in one table to a matching column in another. Many joins can make a program inefficient, and understanding the data can be difficult in a large database with many tables and many joins. I'm not fond of programming with joins and have never used them. I prefer to manage my data relationships through my code.

This data complexity reminds me of another, Indexing, which accompanies relational databases and joins. Imagine a file of 500 phone numbers, each with 200 bytes of associated data, including the phone number. There are methods for accessing this data that are more or less efficient. For example, if you are interested:

- Most inefficient way: If the records are in random order, finding a row using one phone number will require a programmer to read every record and check if each matches the phone number sought. On average, a search will require reading through half the entire file. You will read something like 50,000 bytes of data on average.

- A better way: Another approach is keeping the file in phone number order (or sorted), along with its associated data, which allows you to use the "binary chop" method to locate a specific record quickly. Binary chop, or binary search, is a search algorithm that helps find a particular value in a sorted list or array. The algorithm repeatedly divides the search range in half, eliminating the half that does not contain the target value until the target value is found or the search interval is empty. Think of it as finding the middle of a list, then finding the middle of one half of the list, then a quarter of it, and so on, until you reach the one item you wanted. Binary chop is an efficient way to search for a specific value in a large dataset because it reduces the search space by half at each iteration. While the theory behind this file-searching technique can be complex, the actual code is concise. The number of records read on average is $\log_2(n)+1$, where n is the number of records to be searched. For example, you will need 9+1=10 reads of 200 bytes for this file. You will read more like 2,000 bytes of data, not counting the sorting.

- Fastest way: Indexing involves creating a separate file containing only phone numbers and pointers. The pointers are the computer addresses of where the rest of the associated data is located. By searching through this smaller file (using the binary chop method), the pointer can be found quickly and used to retrieve all associated data in a single read. You will read 220 bytes of data.

There are more nuances, but this is more than enough for now. In the past, a programmer needed to know about database structures, sorting techniques, indexing, and much more to work with large files. However, most of the logic today is built into "canned" routines provided within the programming language. Even so, there is still a lot to learn in I.T.

Of course, at Orange Lake, I also learned much more about the timeshare business, note financing, sales operations systems, satellite communications systems, and more.

CHAPTER 10
MVCI UTILITYMAN (AGAIN)

After leaving Orange Lake, I encountered Steve Burks from Marriott Vacation Club International (MVCI). He was now Vice President of Marketing and Sales under Pete Watzka, SVP Marketing and Sales. He asked me if I would be willing to come back to Marriott. The position offered was Senior Director - Customer Acquisition. My title was a step down from my previous title at MVCI, but MVCI was now a much larger company, and I was more interested in the team and the work than my title. And it would allow me and my wife to remain in Orlando. One area of expansion was the creation of a sizeable outbound telemarketing center.

Call Center Support

Because I had extensive experience in the timeshare industry, I soon became a "Jack of all Trades" for the Marketing Department, like a Utility Infielder in baseball. I started with no staff. I became involved in various projects and tasks, like implementing an extensive Siebel system in the outbound call center. I had extensive reservations and call center experience. MVCI used the Siebel system. The Siebel system was a Customer Relationship Management system for a large call center -- think Sales Management, not unlike Sales Force today. The Siebel system was later bought by Oracle and incorporated into its suite of applications.

For a time, I worked in a second-floor office in MVCI's huge telemarketing center in an industrial park in south Orlando. My office was next to the Telesales department.

When it came to hiring, I had an unfortunate experience happen. I had made an offer on a Friday afternoon to a young man who was so strong that I did something I usually would never do; I told him at the end of the interview that he had the job and should show up on Monday. When he did not show up for work on Monday, I made inquiries and discovered he had committed suicide over the weekend. He had seemed entirely normal. No one could explain his actions.

Most Dangerous Guy in the Room

I was in a meeting with my boss, Steve B., and MVCI's Call Center manager talking to some businessmen from Chicago who were pitching some technology for the MVCI Call Center. These guys were shady, and I knew it immediately. In an earlier visit to their offices in Chicago, in preparation for this meeting, they had offered me stock warrants in their company to get me to use my influence on getting their deal signed.

At one point in the meeting, in response to my asking some pointed questions, the head of the third-party company said, pointing to me, "That guy is the most dangerous person in this room. I would like him to leave." I was happy to oblige and left. They did not get the business in the end, and later I heard the company failed. I usually know a scammer when I hear one.

Privacy and Security

Our division did not have a Privacy Policy expert in the early days of privacy concerns, but Marriott Corporate did. I worked with the corporate VP, studied the subject, and became our division's unofficial Chief Privacy Officer. I attended a seminar on the topic and found that of about 50 attendees, I was the only person in the room who was not an attorney. I knew that when that many attorneys were interested, that litigation would soon be forthcoming. I became as knowledgeable as I could informally. After developing some of the necessary policies, statements, and disclaimers, HR told me to let corporate handle it because he was a lawyer, and the topic needed that expertise. I had learned enough to agree.

I also initiated and led an investigation into the theft of a customer list from the call center. We had 'seeded' the list with fake names and our phone numbers. If we received a call from someone asking for any of those fake names, we would know that someone had stolen the list. This scenario occurred, proving that the list had been stolen.

I traced the data theft's timing to the precise date and time - it was taken over a specific weekend. The list of suspects was narrowed down to a few individuals through key entry to the building and computer logins, and, though we did not pin it down to exactly the one person, we managed to instill in the short list of suspects the fear of potential prosecution. And, with the help of our division attorneys found the list broker who had the list and retrieved it successfully, putting an end to the theft.

Counselor Leadership Training

After these projects, I was assigned to a team of four marketing and sales directors to organize and implement the Counselor Leadership program, a training program for our large call centers on selling skills. The program was to convert our outbound telemarketing agents into Vacation Counselors. Alan Goldstein, National Director of Sales Operations and an experienced Sales Training developer, was our expert and team leader. Vernon Pride was the Los Angeles Call Center Director, and he welcomed the program, so it was very successful. Sadly, I have forgotten the other two developers of the program.

Marketing Database

Another project I was assigned involved the creation of a marketing database. This assignment was made in the early days before Internet marketing. This project was to acquire a variety of customer lists with demographics matching the demographics of those customers who previously bought MVCI timeshares.

I worked extensively with Acxiom. Acxiom is an Arkansas data and technology company specializing in providing customer and audience targeting, data-driven marketing insights, and identity resolution solutions for businesses worldwide. The database was an extensive mailing list that needed enhancing, deduping, and a means to extract subsets of the data for marketing campaigns.

Partner Marketing

One of the more successful assignments was forming a team to expand the Partner Marketing activities that Rob Lamp had been developing alone and extend it as much as possible. Partner Marketing involved working with companies with large membership programs, like American Express (AMEX) or Delta Airlines, and creating an offer for their members to visit an MVCI resort and tour. The Member would pay a portion, like $299, for a three-day, two-night stay at a Marriott Resort. MVCI would pay the Membership Company $50 for each booking. As a result, the partners could offer a valuable benefit for their members and collect more for our appearance in their marketing vehicles than simple advertisers paid.

One beauty of this program was that the Members were already a perfect match for the demographics that MVCI was targeting. This marketing approach is like "Asset Lite" development. By collaborating with partners, you can achieve more collectively than each party could accomplish independently.

Our team consisted of Rob Lamp, who started the program, and others: Jim Reichley, who worked with me at Disney; Travis Griffin; Milton Ferreira; and Kimberly Cusani, as our Administrative Assistant. For a time, we also had on the team: Kristen Weisz, daughter of the company president, and Justin Watzka, son of our SVP of Marketing and Sales. It can be a challenge managing the kids of two bosses. We collaborated on other projects with Carol-Ann Dooley; Rob Hebeler; Mitch Martin, and others. These were some of the most brilliant people I have ever worked with, all hard-working and dedicated to our mission to expand partner marketing into a significant revenue contributor and the other projects we worked on.

The team enhanced tracking within MVCI's marketing systems to accurately measure the amount of revenue generated directly from our initiatives. We built an advanced marketing model in Excel, capturing all potential marketing channels, offers, costs, and the potential of each combination, along with the historical response rate of each campaign. This model then guided us in prioritizing our efforts. By doing so, we were able to optimize and maximize our results.

At times, our methods were too successful. For example, American Expresses' (AMEX's) monthly newsletter had limited slots for third-party advertisements. Our division was paying AMEX more than our hotel division, Marriott Hotels, could afford, which had been enjoying a monthly presence in the newsletter. When we found that we were pushing our parent company out of that media, we backed off, changing to alternate editions to leave room for our hotel arm.

In the first year of the expanded team, we increased Partner Marketing driven sales from $12 million to $75 million. This broader team also assisted in developing a new product concept.

Cabin-in-the-Woods

Our loose band of MVCI Associates formed an unofficial team who wanted to develop a new real estate project called Cabin-in-the-Woods. We worked after hours and on weekends on it. The concept of the project is that it would recapture the experience of family camping but with the comfort of Marriott lodging.

The project would be built on a freshwater lake and constructed around a central lodge. The units would be upscale cabins with all the comforts of home and built with unique features to fit the needs of the various demographics: couples, families with small children, families with teens, and seniors. The resort would be operated around many "camping" and interactive activities run by an Activities Director, including fire pits, story-telling areas, a volleyball pit, a pier with rentable boats for fishing, hiking trails, and other activities. We even identified the first land acquisition for the proposed property. The property would also be highly automated, with self-check-in, property and unit entrance by phone app, and high-speed internet (shut down during activities).

The team worked out all the details and financial models, built a powerful presentation, and scheduled a time to present the proposal to our division's formal Development Committee. The committee included the SVP of Development, Architects, Engineers, Marketing & Sales, and Operations – everyone who would need to sign off. Our team was excited at the opportunity to create a new product.

We received word on the day of the meeting that the company president would attend. That was abnormal, and I became worried.

He came in, sat at the end of the long table, and explained that he did not usually wish to interfere with the committee's activities but that the proposal seemed significant. We presented our proposal, and after the obligatory Q&A, the president took a vote of those on the committee. Every member voted in favor except two. One said he voted against it only because he thought the project should proceed more quickly than we had proposed but was otherwise in favor. The other no-vote came from someone the president had spoken with before the meeting, as he gave no specific reason for his objection. Then the president announced that he usually did not interfere with the committee's decisions, but in this case, he would have to do so and turn down the project. Everyone was aghast except for the CEO and the other person who voted no. After all our work to develop the product, it never left the starting gate. I was very disappointed.

After the meeting, I learned from some friendly committee members that they believed the president was afraid of approving another project that might fail. He had been the "parent" of two major projects, Horizon Resort and Ritz Carlton Club, both of which floundered. I knew about those two projects during planning and had tried to intervene before they were far along because I recognized flaws in both, but my input was ignored. And in the end, both projects failed as I expected.

Here is why the Horizon product failed. The Horizon Resort was a lower-tiered product than the MVCI Resorts product. Still, the company insisted on sourcing customer prospects as they did for the MVCI product by telemarketing or capturing in-house hotel guests, whose demographics were too high for the product. Like other timeshare developers, they needed to pull prospects in via OPC (Off-Premises Contacts) locations. History has shown that people who can be talked into taking a tour off the street for an incentive worth about $150 are the right demographics and have the propensity to buy a "mid-level" product. Driving the same demographics as MVCI did with tour packages did not differentiate the product enough from the MVCI product. It might have worked as a "Drop Product," meaning if a customer does not buy MVCI because of price, then sales could have shown them the very nice but less expensive Horizon product, but MVCI did not use Drop Products then.

The Ritz Carlton Club was less successful than initially hoped. It did not include a full range of Exchange options. In trying to reinforce its exclusivity, Exchanges were limited to the Club only. Also, the product was designed as a Fractional Product but did not fit the success parameters for a fractional I had learned in Missouri – being within driving distance. It lacked convenience for distant luxury customers. So, with two failures under his belt, no way the president was going to risk a third.

We began working on the Partner program for the following year when a recruiter contacted me. Then I received an offer I could not refuse.

What I learned at MVCI

Throughout my career, I have worked with a diverse group of individuals in both start-up organizations and larger companies like United Airlines and Marriott Vacation Club. Valuing open-mindedness, I have always sought alignment with my superiors and peers, fostering an environment of understanding and collaboration that allows me to build strong relationships with like-minded individuals.

One crucial lesson I have learned is the significance of collaborating with supportive and forward-thinking superiors. I thrive when surrounded by leaders who embrace innovation and encourage calculated risks. This approach has enabled me to make significant contributions to various start-up organizations, where camaraderie and honest communication played pivotal roles in our collective growth.

Although I faced challenges in larger organizations like United Airlines and Marriott Vacation Club, I approached these experiences as valuable learning opportunities that have shaped my professional growth. As I progressed to the director level, I discovered my ability to excel in environments where company politics are minimized. Moreover, I found my best performance overseeing well-defined areas, reporting directly to company owners or supervisors who allowed me autonomy, and fostering a sense of independence and open communication.

Throughout my journey, I have come to appreciate the importance of surrounding myself with individuals who prioritize transparency and teamwork, regardless of the organization's size. By finding the right fit and minimizing politics, I have made meaningful contributions to the success of the companies I have been a part of.

I did not advance my technical expertise much while working at MVCI, but I was able to add value to some projects using my prior experiences in systems development and project management. I did contribute using my technical skills with the Siebel installation and in helping to build a marketing database.

Basketball

While there, to help build my peer relationships, I organized a twice-weekly basketball match. We always had ten players or more and often had to rotate players when more than ten showed up. The ten players had a few reliable regulars, like me. On any given week, the participants changed by about 50%. I ran this informal event for several years. Initially, we played at a small court surrounded by tall chain-link fencing at a nearby hospital, which they kindly allowed. We called that the "Cage." After the first year, the hospital decided they needed more parking spaces and removed the Cage. So, we moved over to a court at one of our resorts.

Because the court was an amenity for the resort guests, we avoided interfering with our customers' vacations. If guests came while we were playing, we would offer to include them in our game, even when the guests were kids or teens. Fortunately, the guests always accepted our offer, and no one complained. The company would have kicked us off quickly and permanently if they had.

I remember a time when we included two teen girls in our game. As we learned later, these basketball players, around 14 years old, were starters on a state championship team in their hometown in Nebraska. Their skills, agility, and comprehension of the game far exceeded our expectations, especially given their age. They consistently landed three-pointers, maneuvered around the court effortlessly, and were tough to guard. Their participation added an extra layer of challenge to the game.

When it was time to leave my employment at MVCI, I organized one last game and invited all my peers who had previously participated. It happened that my very athletic son was in town, so I asked him to play. However, we conspired to hide his identity from the others, planning instead for him to pose as a resort guest. As our protocol was to invite guests to join us, he was warmly integrated into my opposing team. Our plan was for him to deliberately disrupt his team's flow by occasionally passing the ball to me instead of his teammates or blocking his teammate's shot.

The first time he passed the ball to me, his team was livid but had to be restrained and polite to a Marriott "guest." The second time he did it, a player on his team got way too angry. So, I had to break my son's cover. I told them, "I would like to introduce this guest to you. He is my son."

I told them that they had just been "punked!" All those there, twenty in number, went into convulsions laughing. All except the most incensed one. I do not think he ever got over it. It was a memorable way to leave the company,

MVCI would convert to a Point System after I left, and for that, they could have used my help, but it could not offer me enough to keep me from heading off to a new company when the opportunity arose.

It is a natural evolution to move from Programming to managing programmers, into project management, and then to general management. If you have been programming for a company for a long time, you know the business more intimately than most general businesspeople. Because every business function gets automated or touched by technology at some point, those in technology see the inner workings of the business. Being integral to the company's technology, you become intimately involved. For this reason, many CIOs become CEOs. In my case, I had been heavily involved in automating every functional area of the timeshare industry. It was then natural for me to eventually lead a company.

CHAPTER 11

GOLD KEY RESORTS

A recruiter contacted me on behalf of a multi-resort company in Virginia Beach, VA. The position was President of Gold Key Resorts. The primary mission was to optimize all the resorts' processes, especially those surrounding marketing and sales. This assignment involved enhancing both automated systems and manual processes. Throughout my career, I have discovered that improving manual procedures adheres to the same principles as enhancing automated ones: organize, simplify, streamline, optimize, automate when feasible, and train people thoroughly on the procedures.

This job offered a marvelous opportunity to utilize my full range of experience while also allowing me to grow professionally. I was given complete liberty to run the Marketing and Sales operations, with one exception. I endured intrusive oversight by two "spies" in accounting and a VP of Sales who played golf with the Owner almost every weekend and liberally shared whatever my team was doing, but not always in a constructive way.

I recall a meeting I arranged with the Sales Team to develop our goals for the first year. Typically, I would introduce myself to a new team by presenting a simplified organization chart with vacant boxes, encouraging them to anticipate how I might populate it. When I eventually filled it, I placed them in the top box and myself at the bottom, illustrating that my role was to serve their needs, provide the resources they needed and assist as needed. Initially, I could tell from his body language that the VP of Sales did not care for being in the box second from the bottom. This method was well-received and accurately represented my perspective on my role and his.

I then asked the team what they thought their goals should be for the following year. As I started this exercise, the VP of Sales took me aside and said that my approach would be a fiasco and that I should stop immediately. I did not. As the exercise continued and we honed down the goals, it became clear that the team, as I often found them to do, set more challenging goals for themselves than the VP of Sales would have set or thought possible. He was man enough to apologize when the meeting ended and exclaim how astounded he was by how well it worked out.

The Owner, Bruce Thompson, was a brilliant man, a visionary, and a man of his word, all traits I respect. He embodied a principle I have tried to live by that ethics are essential in business. Bruce was a stern taskmaster but always honorable, fair, up-front, and ethical.

For a time, his son, Josh, worked as my head of OPC marketing. Josh was great – personable, able to diffuse any tension that arose with humor, and became a good friend. In addition, he taught my two grandsons, my son, and me how to surf.

A few weeks after I settled into our new condo near Chesapeake Bay with my wife and loyal lab retriever, I needed a haircut. I asked our CFO, a neighbor, for his recommendation. He directed me to a nearby barber shop.

The barber shop did not take appointments; you just waited your turn. I took my place in one of the seats along the wall until the fourth barber chair finally opened. I sat down and gave the barber my usual instructions on how to cut my hair. However, once he was finished and I saw myself in the mirror, I was shocked at how short he had cut it!

The next day, I decided to share my frustration about my haircut with the CFO, only to be met with an unexpected response. "Oh, did you have the fourth chair barber cut your hair? I should have warned you. He cuts the hair for all the Navy Seals in the area." I experienced my first and only military haircut.

We had some memorable experiences at one of our annual company picnics. The first was a touch football game, where Josh was the quarterback on our team, and I was the right wideout. Everyone playing was young except me. On the first play, I told Josh not to throw the ball to me. I limped out on a weak down-and-out pattern. On the next play, I told Josh to be ready to hit me with a pass. I started out feigning the same weak pass pattern and then broke downfield. Josh's pass was spot on, and I scored a touchdown. I heard the other team's captain yell, "Who had that guy?"

On our next possession, I told Josh that I would run to the right front corner of the end zone, and he should pass the ball so it was just out of bounds, where it would be hard to defend. He again hit it perfectly. I caught the ball for a second touchdown. After that, the other team put a speedy athlete to cover me, and I did not score again, but we won.

The other picnic activity consisted of throwing whipped-cream-covered pies at managers, who were to take turns sticking their heads through an opening in a playfully decorated and colorfully painted plywood board. It seemed innocent, although I never thought degrading each other was a good idea. Josh went first and learned on the first throw that the person who bought the pies had bought frozen pies. Josh ended up with a bloody nose, which ended that activity.

Josh was a great help to me and did a great job running his OPC department. Sadly, shortly after our first year working together, he contracted ALS just after his marriage. Even more tragically, he found out that his wife was pregnant around the same time he was diagnosed.

He contracted the illness right at the beginning of the summer, our busy season. It felt wrong to hire his replacement while he was starting his battle. So, I took it upon myself to do his job and mine for the summer. With the help of his assistant, Leanne Tobin, I managed all his OPC sites. On weekends, I rode a bicycle up and down Atlantic Avenue overseeing and motivating his team. Leanne was incredibly helpful in managing the office operations, providing invaluable assistance that I could never fully reciprocate. We succeeded in maintaining Josh's expected production levels for him that summer.

Josh fought the disease bravely for years, even having a second child. Unfortunately, he did not live long enough to see a cure invented, although he and his family worked and still work tirelessly to raise money for and support research into ALS. His loss was deeply felt by his father, his mother, his family, his work colleagues, the community, and personally by me. It was indeed a significant loss for all of us. Meanwhile, I had to keep up with my job.

Systems

Goldkey Resorts' computer systems managed Off Premise Contact (OPC) tour generation, outbound telemarketing, and sales contracts. We had two programmers because it is never a good idea only to have one programmer – no backup if one were to leave or become ill.

James Sacra was our primary programmer and an exceptional one at that. He was known for his speed, efficiency, and solid coding skills. His contributions included the following:

1. Redesigned our primary marketing management system.

2. Developed and managed the company's automated booking process -- from inventory search to purchase and confirmation process.

3. Developed a resource-tracking application for the resort division.

4. Introduced Online Analytic Processing (OLAP) for our marketing data using MS Analysis Services.

5. Designed key performance reports based on accounting requirements using Crystal Reports.[5]

When I initially joined the company, our small IT team was not using VB.NET. VB.NET (Visual Basic .NET) is a programming language and development environment that is part of the Microsoft .NET platform. It is an evolution of the classic Visual Basic (VB) programming language, designed to be fully object-oriented and to take advantage of the features provided by the .NET Framework.

VB.NET is used primarily for building Windows applications, web applications, and web services. It provides a user-friendly and approachable syntax, making it a popular choice for beginners and developers who are transitioning from other programming languages. James expressed his interest in learning and using VB.NET and promised to do so quickly. Despite needing a critical system completed, I agreed to let him dedicate the time to learn VB.NET, with the condition that the system had to be operational within three weeks. Remarkably, he managed to accomplish both tasks within that timeframe. Once he became proficient in VB.NET, he showcased an exceptional ability to produce high-quality code as efficiently as the most skilled programmers I've encountered. His expertise also proved invaluable in streamlining our marketing efforts.

Marketing Expansion

To expand our partner marketing, I hired Rob Lamp from MVCI to collaborate with me in Virginia Beach since he had crested in what he could do with Partner Marketing at MVCI. As VP of Marketing, he helped to keep the sales lines filled with tours and cleaned up the telemarketing operation.

Leo Ortiz was another major contributor. He managed our guest Welcome Center, which served as the first check-in point for our tour guests upon their arrival in town.

Streamlining this operation required a sizable effort by Leo, Rob, and me, but we were ultimately successful.

Redskins Beach Blitz

My other activities at Gold Key were not systems-related but primarily marketing projects – a massive marketing partnership deal with the Washington Redskins (now the Commanders), the opening of new Offsite (OPC) marketing locations, including one which was an aquarium, research for possible development in the Outer Banks and a Ski Location, and the expansion of outbound package sales.

The Washington Redskins partnership was a unique opportunity that I took on, initially without telling my boss, as I was not sure how he would take it. The city was having a weekend event with the Washington Redskins football team called the Beach Blitz, where all the players and management would come to Virginia Beach for a weekend with their families.

A marketing executive from the team came with the Virginia Beach City Administrator to see me, hoping to reserve all the rooms in our Hilton Hotel. Their visit was far enough in advance that we had only a few bookings in the hotel, and we could move them to our timeshare condominiums as upgrades so that we could do it. The city planned to pay for the rooms. However, I proposed instead to provide them with the hotel rooms for free, which would free up the money they were expecting to pay, $50,000 or so. The city could then use those funds to promote the event further. The city's plans included having an experience equivalent to the NFL Experience operating in the convention center, and the team would have a family function on the beach. But I wanted something in return for our contribution.

I traded the hotel rooms to the Redskins for permission to promote our resorts in their stadium. They ultimately agreed to allow us to have a booth inside the stadium during all the games that season, work the parking lot with mini-vac promotions, attend their training camps, and include eight-game tickets and lots of merchandise for promotional purposes for each game, plus an invitation to see a game in a skybox. The tickets were especially valuable because the stadium was sold out every game, and they had a waiting list of tens of thousands of fans wanting tickets. The city also permitted us to hand out lead slips and promotional flyers in the convention center and to everyone waiting for their turn to go in.

Even though the hotel rooms we provided were for one weekend this year, I made a deal for two years. This partnership represented a significant marketing opportunity for our company.

Once I had a contract for the deal, I took it to our next executive meeting. Because the deal was done without prior permission, I meekly asked Bruce what he thought about the Redskins. He shared his childhood experience of growing up with his single mother, during which they would watch the Redskins together every Sunday. He was ecstatic about the deal we had agreed upon.

The only hitch with the agreement was the wife of Bruce's silent partner, who called me demanding that I give her the game tickets we had received for promotional purposes only.

I explained that the tickets had to be used for promotions of our resorts and were not for use by company personnel. She was livid, and I soon got a second call from her husband, again insisting. All I could say was that Bruce had his tickets; if they called him, he might provide them with some. I held my ground on their demands where another person would have caved in and, in doing so, jeopardized our excellent relationship with the Redskins.

The attendance in the first year at the Virginia Beach event, across all venues, exceeded 10,000 fans – a tremendous success for the city. In terms of Gold Key's investment, a comprehensive analysis conducted by our less-than-friendly accountants after the event confirmed that we generated revenue and profit significantly surpassing the cost of hotel rooms and game staffing. The overall deal we struck served as an example of the potential of creative marketing through partner relationships, a skill I had now mastered.

My Favorite Sales Meeting

We had three sales teams or lines, one for each type of customer – one for OPC tours and Mini-vacation tours, and one for In-house guests, those vacationers already staying in our resorts or our Hilton Hotel. One of my duties was to kick off periodic sales meetings for these lines. Sometimes it was the Inhouse sales team, and other times it was our two larger General Sales teams.

Our Inhouse Sales team was unusual for the timeshare industry in general and especially for Virginia Beach, which had a history of racial tension. The team was surprisingly composed of all African American men, about 12 in number. They were led by Mary Reinhart, who was not African American, and one of the two best timeshare sales managers I had ever met. Her team had the highest closing rates and sales volume of any sales team I ever knew. The whole team was terrific. However, any group of salespeople can be unruly, so I often had to go with the flow. I always tried to open my pep talk with some comment aimed at connecting with them.

In one meeting, I had recently completed my DNA genetic testing and sought to use the findings as a connection point with the team. I shared my amazement about how my ancestry, based on part of my DNA tracing back to our collective origins in Africa, made me realize that we are all historically linked. In response, one of the salesmen retorted, "Mister Patton, one thing you ain't is one of the brothers!" This comment sent the room into fits of laughter, including myself. I was left speechless but continued, changing the subject to their recent commendable performance.

What I Learned

I learned how a master decision-maker drives an organization – Bruce Thompson was masterful and a serial entrepreneur. He controlled his environment. He held leases on every possible OPC location along the west side of Atlantic Avenue (the city outlawed them on the east side). His real estate developments were artfully designed. For example, he created a highly successful restaurant in a perfectly designed Hilton Hotel, made through an innovative public-private deal with the City of Virginia Beach, then voluntarily donated the land next to the hotel to the city for a small outdoor entertainment venue. He was the perfect balance between quality, cooperation, and profitability. From him, I learned to be decisive and trust my gut.

I also learned how to impress a new hire. In my second week on the job, I was told by my wife that I was invited to join some people with her after work at the Hilton. As I walked into the Hilton, my wife was waiting. She led me into the elevator with one of the hotel's managers already there. My wife tried to press the button for the private level on the top level, but it wouldn't light up. The manager accommodated her and used her key card. My first reaction was that the manager's behavior could have been a better security practice than letting unknown people onto the secure top level.

When we got off, my wife led me onto the top-level infinity pool deck. As I walked out into the open, I was greeted with a loud "Surprise." Bruce, my boss, had organized a surprise "Book Signing." They had purchased over 100 copies of my **Colt Terry, Green Beret** biography and set up a table where I could autograph a copy for every manager and executive in the company. It was a great way to meet every management person in the company and for them to meet me. It was a brilliant idea by Bruce. It touched me deeply, even though I didn't remember most of those I met.

After two years at Gold Key, which I immensely valued, my time there ended. The owners wanted to go in another direction. I had some unfinished business but was able to leave Bruce with a game plan for the upcoming year.

My wife and I enjoyed living in Virginia Beach. We liked the social life, the availability of culture, especially the early history of America, the bay views from our balcony, and living four minutes from the beach. We would have stayed if another opportunity had arisen.

CHAPTER 12
GRAND CROWNE RESORTS

Branson, MO, and Pigeon Forge, TN

After leaving Gold Key Resorts, I was connected through an intermediary with Randy Black, a colleague I had known at MVCI. He had led MVCI's development of a "mega-call-center" in Tampa that was state-of-the-art and a "work-of-art." He converted a large part of a long-neglected shopping mall with ample parking, a location close to a university, and a food court. Unfortunately, after the "showcase" center was up and operating, the company had a change of heart and closed it immediately after it opened.

Like me, he was a free agent and had been engaged as COO by C.J. Perme, owner of arguably the largest timeshare resort in Branson, Grand Crowne Resorts. Randy hired me as his V.P. of Marketing. It was 2008 and again, another hard economic time with the subprime mortgage crisis and the crash of the Housing market. During the global economic downturn, banks ceased lending, and development stopped worldwide, not just for timeshares. Interestingly, this climate provided opportunity and proved beneficial for us. Larger timeshare companies, like Bluegreen and Westgate, were downsizing their operations in Branson, abandoning their Off Premises Contact (OPC) locations. This abandonment created an opportunity for us to step in, taking over these abandoned locations at lower rents.

When we first arrived, we learned that the company had an agreement with Resort Condominiums International (RCI) for exchanges in Branson and with Interval International (II) for exchanges in Pigeon Forge. This arrangement was unworkable, as owners in one location could not exchange to the other. We needed to pick one company first. We invited them both to talk to us about our business. RCI sent some information, while Craig Nash, CEO of II, came in person to pitch us. That personal touch made the difference, so we went with II.

Next, we had to get out of our RCI contract. Randy and I went to the next ARDA conference and scheduled a meeting with RCI. I prepared a case for why RCI should let us out of the contract, which exchange companies were known never to do. We planned for Randy to pitch the case, as he was a preeminent salesman. Between my case and Randy's skills, we successfully received RCI's agreement to let us out of the contract.

Randy managed sales and overall operations at both resort locations, while I likewise managed the OPC locations, sales operations, and marketing functions. During this period, he and I also oversaw the operator of a third-party telemarketing operation in Springfield, MO.

The owner also owned a large theatre, like those all over Branson with Country Western entertainers, except this one was vacant. Randy and I tried to revive it by developing a show around a stage magician. He was good, but without investment in billboards and advertising, it was not feasible in our economic climate and much-reduced visitor traffic in the town. So, we focused on expanding our network of marketing locations.

However, we did not stop with just Branson. We saw the potential for growth in Pigeon Forge and started expanding our business there. The owner had started a sales operation there but had yet to build a model cabin. Our initial focus was setting up a Sales Office and identifying marketing locations. We were able to hire Bob Wilcoxon, who had led sales for Bluegreen before they closed their sales operation due to the poor economy; he was a fabulous sales leader, built a superb sales team, hired many of his prior Bluegreen team, and made our success there. Eventually, he was given authority over all three of our sales operations. The timeshare units were cabins and needed a model built for sales to begin, so C.J. started on that.

The situation in Pigeon Forge mirrored Branson. Major timeshare developers were closing, leaving OPC locations vacant, which we promptly occupied at a cost less than our competitors had been paying. In Pigeon Forge, we acquired a prime visitors center on the entrance road as an OPC location – a significant victory. We also secured more OPC locations in Branson, including in the lobby of the Hilton Hotel. Despite the decrease in travelers in both destinations, the traffic was substantial enough to keep our sales operations busy.

Our resort company was able to sustain its momentum in these challenging economic times thanks to our local banking relationships, competitors downsizing, and a strong sales team in both destinations. Despite the economic climate, we continued to grow.

After living for a few winter months in a company unit, my wife and I moved to our own rental home when tourist traffic began to build for the summer. We were lucky to find a small house on the local golf course. It was a decent base, given that I was traveling some.

Unfortunately, the distance between Branson and Pigeon Forge was such that we had to fly back and forth by two-engine aircraft with pilots that made me uncomfortable when they risked flying in unpredictable weather around the Smokey Mountains.

Another Near-Death Experience

After flying for years on very safe commercial airlines, and having no fear, flying in small aircraft with unknown and potentially questionable maintenance conditions still made me nervous. The company had access to two different two-engine turboprop planes, a smaller one and a larger one. Both were kept at a small private airport near Branson, not Branson Airport, which had yet to be built. We always departed precisely at dawn, when the sky showed light, to maximize our working time on the other end.

The pilots varied from a list of two. One was an extreme conspiracist, who paid no income taxes, spouted extremist views, and hated the government. He made me wonder what he thought about FAA aircraft maintenance regulations. I knew less about the other but suspected he was similarly inclined based on his decision-making.

The planes always had a few spare seats, so we had room to stretch out. The flights were usually smooth, long, and tedious until we reached the Smokies. The small airport in Pigeon Forge was subject to frequent crosswinds and buffeting on landings and take-offs. The owner, C.J., could have been more talkative and would usually sleep, I would try to get some work done, and Randy kept our conversation lively, often debating politics with me. Riding together for seven hours round-trip, we ran down most topics. Fortunately, the pilot was too far up front and on headsets, so he rarely conversed.

My most harrowing experience occurred on a return flight when we flew into massive, black clouds that were too high to fly over and too wide to fly around. Our stubborn pilot passed through them directly instead of seeking an alternative airport. He flew headlong into dark black clouds with flashing lightning surrounding us and violent turbulence shaking the plane. I worried about the aircraft disintegrating. I could do nothing more than grit my teeth. The ordeal seemed to last forever, but eventually, we emerged from the storm just as we reached Branson.

My second harrowing flight was when Randy and I flew to Pigeon Forge with the President and EVP of MasterCorp, a large, professional Property Management company. We planned to show them our development so they could give us a quote for providing their services there. MasterCorp's aircraft was an EMBRAER SA EMB-550. This plane is a two-engine small jet that was roomy and comfortable inside. When the FAA approved the EMBRAER SA EMB-550, they wrote, "[These airplanes] have a novel, even unusual design, and feature a vision system that displays video imagery on the head-up display." They also said at the time that "These aircraft required special airworthiness regulations with additional safety standards that the FAA considers necessary to establish a level of safety equivalent to that established by the existing airworthiness standards."[8] I could see by looking at the airplane that its design was unusual and appeared unstable.

My concerns were confirmed on our final approach to the Pigeon Forge single-runway airport early that morning. The winds were blowing hard across the runway, not from our front, which is what pilots usually want. The winds were gusting strongly too. The plane was bouncing all over the place. I had often experienced aircraft landing in heavy crosswinds and the pilot "crabbing" or "sidling" the aircraft onto the runway, but this was worse than any experience.

Finally, just as we were about to touchdown, an even stronger wind gust hit the side of the aircraft and tipped it over to the left. I watched as the left wingtip came closer and closer to touching the tarmac. If the wing touched the ground, the plane would spin, crash land, and possibly catch fire. Thankfully, the pilot regained control at the last moment. We landed nearly sideways and with a jolt but avoided an accident - an encounter too close for my comfort.

Fortunately, C.J. flew in later that day on one of our company's planes so that we could return to Branson in our airplane. I never thought I would feel safer in our company airplane than in a modern two-engine jet, but this day I did.

Biloxi MS

One of our goals as a company was to open a third location. We met the owner of a block of condominium units in a converted motel property in Biloxi. The property already had several attractive amenities – a large swimming pool, gulf views and beach within walking distance, and a nearby casino. It also had room to build more units on an adjacent lot. The condos were roomy and nicely refurbished. They met our standard of being in "like new" condition.

However, individuals owned and used the units as residences or second homes. Because of the housing financial crisis, the resale prices for the units were going for less than half of what the owners paid for them, if they could be sold at all. Every owner stood to lose more than half of their original investment, and many had mortgages far greater than what the units were worth, so a local bank cared about what we might do too. The idea of converting them to timeshares provided the only hope that the owners might recover their initial investment under the right circumstances.

This project would be another new "asset light" development form for Grand Crowne. Buyers would be buying into a timeshare company with a system of three resort destinations. We would market and sell the units as timeshare weeks, keeping a commission for Grand Crowne.

To form the deal depended on the head of the Condominium Owners Association, our contact, Bernie Burkholder, convincing at least half of the owners to convert to timeshare. He owned eight units and had his own "skin in the game." Long story short, he miraculously convinced 100% of the owners to convert to a timeshare resort. We also needed the local bank's approval, which held the owner's mortgages. Bernie and I met several times with the bank to get their approval, which Bernie obtained after several meetings.

For marketing locations, I negotiated a deal with a local casino to open a tour desk adjacent to their gambling floor. This arrangement proved highly successful. I also worked on signing up a tour desk in another larger casino and one at Hard Rock's casino. The potential for expanded marketing with no competitors in town was very promising.

We had arranged a tour desk in the Biloxi airport for a time, right at the bottom of the arrival escalator. This deal would have been an enormous success, but just as we were about to open it, the company that controlled all the advertising in the airport, Clear Channel, accused the airport of breaching its contract with them. The airport backed off unless we could strike a deal with Clear Channel, who refused. We then altered our Branson Tour Packages to include a secondary offer to visit Biloxi after they visit Branson. Finally, we had enough tours to run a small sales line and gradually sell some units. However, if it was not one thing, it was another.

We rented and connected two large mobile home trailers for a sales center, which we had also done in Pigeon Forge. This approach was straightforward, quick to implement, and less expensive than permanent construction. In this case, we only had a section of a parking lot, so a permanent sales center or clubhouse was not feasible.

Once sales began, sales were slow but steady, so the plan was working. However, after about six months, the BP Oil spill happened, which closed all the beaches along the gulf and brought tourism to a hard stop. During recovery, our operation bumped along for a while. But with the economy poor and Biloxi's potential threatened, the Grand Crowne owner decided to trim staff, which included me. So, it was again time for me to move on. So, my wife and I drove from Branson back to Orlando. I knew I could submit a claim for lost wages with the fund BP set up, which I did. I was surprised about a year later when one day, a check arrived for a year's worth of my salary. Even though I had alerted the company's attorney, I do not think the company or anyone else who had been impacted even thought to apply.

Firsts

A first for me was being awarded Interval International's Chairman Club for being one of the top 25 member-producing companies in the II world in the face of the worst possible economy. This award was a testament to our marketing and sales teams in Branson. Especially my right-hand marketing manager, Stephanie Smith, was endlessly helpful in marketing at all three locations. And we achieved this sales level during one of the worst economic times in America.

Another first was using all my timeshare experience to open complete marketing operations in two new locations and successfully managing three OPC managers in widely dispersed locations. Managing OPC managers, like salespeople, can be like herding goats or chickens.

CHAPTER 13
RETURN TO CONSULTING

Hilton

After returning to Orlando, I took up consulting again for a time. I met with Mark Wang, who had just become President of the Hilton Grand Vacations Club (HGVC). Mark had done a bang-up job of developing Hawaii resorts for the previous president of 13 years, Antoine Dagot, who had built up HGVC after it was founded in 1992 by Edwin H. "Ed" McMullen, Sr.[9]

Mark took me on for a few months while transitioning into his new role. I had hoped to find a fit somewhere in his organization, but no opportunity that fit my skills arose. It was 2009, and I was 67 years old. So, the time had come for me to retire, or so I thought.

Ritz Carlton Club Again

Pete Watzka, for whom I worked previously at MVCI, referred an opportunity to me. The Homeowners (HOA) Association's Board Chairman from the Bachelor's Gulch Ritz Carlton Club (RCC) resort wanted advice on changing management companies. All the RCC properties were now in the hands of the property Boards and no longer owned by MVCI. However, the RCC Management Company, which was part of MVCI, still managed this property. The Board wanted to consider changing management companies from the Ritz Carlton Club to one of several other possibilities. They asked me to run a Request for Proposal (RFP) process to gather the information needed for the Board to choose another management company. I successfully ran an RFP process at RCI and Disney, so I knew what to do.

Because I had worked for MVCI and I felt that the MVCI's President might object to me using any knowledge I might have from my time at MVCI, I made clear to the Chairman that I would be doing the RFP process very thoroughly and as objectively as possible, which I did. I told their Board in advance that I would not make any recommendations but would do my best to provide them with the necessary information so they could decide. In the end, the Board made a change to Exclusive Resorts.

I would later be interviewed by another RCC property entertaining the exact change. However, they had more complexities than BG, so we mutually decided not to pursue an RFP.

My experiences with these Boards confirmed my original contention that the product could have been designed to address some of the later concerns. However, with all my experience, they never asked my opinion, even though I had tried a few times to express it.

Lesson Learned

In business, always listen broadly and intensely – be like Eddie Carlson and listen to everyone. As a result, many mistakes can be avoided, problems fixed, and better products produced.

CHAPTER 14
VACAYHOME CONNECT

Vacation Rental Distribution

In late 2013, I met and gave some advice to an entrepreneur, Sunil Aluvila. He contacted me about joining him in his start-up. Initially, his company was called VacayStay Connect, but after a minor conflict with another company about the name, he renamed it VacayHome Connect (VHC). The company's product is an inventory distribution system combined with support services – a Software as a Service (SAAS) product. Inventory, in this case, means the availability of a vacation rental home to be booked, like booking a hotel, but in this case, a professionally managed vacation rental home.

Vacation inventory is perishable, meaning if it is not booked, it goes wasted, like airplane seats. It also has many rules around its booking, such as cancellation rules, deposit rules, applicable taxes, fees, and other details. To make this kind of information available and bookable online requires people and processes to collect that information into a central database and make it available to the public by feeding it to online booking engines, like Expedia.com, VRBO.com, and Booking.com. This SAAS distribution is what VacayHome Connect (VHC) does. It does not sell directly to vacationers, only to retail outlets that do.

VHC's business depends on suppliers and distributors. VHC does for its suppliers as Amadeus does for global airlines, and Pegasus does (as Thisco did) for hotels. If you are unfamiliar with these companies, they are central switching hubs for their respective kinds of inventory.

Back into Sales

VacayHome Connect targets two types of suppliers or inventory sources. One type is timeshare resorts. Initially, we made overtures to all the major timeshare companies, with mixed results. The benefit of this program to timeshare resorts included generating revenue for empty units and getting highly qualified renters with high potential to buy memberships for the developers' Clubs. My role included representing the company across the timeshare industry and selling them on distributing their rental inventory through our channels. We secured early agreements with Bluegreen and Diamond Resorts, demonstrating the effectiveness of VHC for timeshare developers. However, we encountered difficulties with major brands because their parent hotel corporations preferred to manage distribution through their hotel reservation systems, even though none were designed to handle rental home inventory. Consequently, they have yet to effectively distribute this inventory through their hotel systems, if at all.

The second supplier type was professional rental homes, condos, and cabin managers. From inception, Myles Snyderman and I served as the entire sales team. We signed up thousands of rental homes, with Myles signing up the most. I also met with a Japanese rental property company that would later sign up.

At the time of writing, the company actively seeks new channels to distribute the large amount of inventory it has aggregated beyond just the online rental home platforms. The primary current initiative involves breaking into the incentives and loyalty program markets through 'private partnerships.' This initiative would provide a web link or webpage where incentive, travel, timeshare, or loyalty club members can use their accumulated points to book vacation rental homes. VHC also can facilitate access via an API (Application Programming Interface) - a method that enables a direct electronic connection between VHC's computer and theirs. The Private Partner would have to do some programming to display the information, so Weblink is the fastest and easiest way to provide rental home access to members.

Capital constraints and Covid impacts have impeded the company's growth, but it is still working hard to grow. Its potential remains strong, and efforts to grow will continue.

CHAPTER 15

FINALLY, RETIREMENT

After Gold Key, I did a bit of consulting for Hilton Grand Vacations, the Ritz Carlton Resort at Bachelor's Gulch, CO, and VacayHome Connect, a Software-as-a-Service distributor of vacation home rental inventory, with which I continue to have a relationship.

Finally, the time came when I could pursue some of my original ideas. I never retired completely but reprioritized my interests. Once I did that, I took several online classes in current programming languages and tools, including PHP, HTML, CSS, JAVASCRIPT, MYSQLi, Filezilla, cPanel, and others. I have rented a virtual private server to control the security of my websites.

The coding of innovative applications is an art form. As oil painting and sculpture are art forms, designing and coding a computer program is similarly creative. I now describe myself as a 'Code Artist®' on my website, chuckpatton.com.

As part of my art, I began working on a couple of my original ideas:

Job Leveling

The first system that I wanted to develop was a Job Leveling system. Job Leveling is a term I use to encapsulate the methods we used to create the manual processes Jim Kent called his "Management Information System" when I worked in Food Service for United Airlines. This system is timeless, and the concepts deserve to be preserved. So, I wrote up a website, database, and background processing to capture and automate these concepts.

Here is how it works: Determining the appropriate level for new positions in a company with various job levels (such as manager, director, and vice president) is essential, but consistently assigning titles can be challenging. The job leveling system is designed to help by analyzing and classifying jobs into levels based on differences in core characteristics such as responsibilities, authorities, skills, knowledge requirements, and pay ranges. It can then determine the correct level for a new position based on where it ranks on those same attributes. The job leveling system uses multiple regression and neural nets (used in AI) to analyze the distinguishing aspects of each job and level. To access this system, please visit jobleveling.com.

If you are interested in the difference between Multiple Regression and Artificial Intelligence (AI), this is how I see them. Multiple regression and neural nets are techniques used for analyzing data and making predictions, but they differ in how they approach the problem.

Here is how the Multiple Regression works, if you are interested:

Multiple regression is a statistical method that examines the relationship between one dependent variable and one or more independent variables. It tries to find the best-fit line or curve that describes the relationship between the variables. The resulting formula, usually a quadratic equation, is created by building a matrix (think spreadsheet) of sample data and then using a statistical computation to determine an equation. That equation can then produce a new result by using new inputs.

A Multiple Regression result formula will likely change to some degree if given a new sample data set. Also, the formula should get more precise with increasing sample data. With an increasing number of independent variables, however, adding new factors that drive the dependent variable result may or may not make the prediction more.

Here is how Neural Nets work, if still interested:

A Neural Net, on the other hand, is a type of machine-learning algorithm inspired by the structure and function of the human brain. It consists of multiple layers of interconnected nodes, each of which processes a small part of the data and passes it on to the next layer. The algorithm adjusts the weights of the connections between nodes to learn and make predictions. The algorithm needs to "learn" by being fed training data. Once trained, the resulting model can produce results even for combinations it has not seen before.

In simpler terms, Multiple Regression is a statistical approach that tries to find the best equation that relates the input variables to the output variable, while a Neural Net sees everything as a network of nodes and their interconnections. It is an artificial intelligence model that learns and makes predictions by adjusting the weights of interconnected nodes. It can adjust its weights with new learning and thereby be dynamic.

As I did, you can use both and then visually compare the results to see which better fits your expectations.

DIY Timeshare Exchange and Rentals

Another idea was to provide a much less expensive means for timeshare owners who can arrange their exchanges directly with their resort to connect with other similarly capable owners. My idea was to avoid the matching processes used by the two major Timeshare Exchange companies (RCI and II) and provide the equivalent of Want Ads. The want ads would allow timeshare owners to connect with others to arrange their one-to-one swaps. This system will save timeshare owners hundreds of dollars annually and provide a place to put their weeks (or days) up for rent, accessible to even non-timeshare owners. In addition, this website can tap into a new source of vacation rentals for travelers and put renters and self-exchangers into resorts that can incentivize them to tour and sell those travelers. I limited the website to the U.S. and am still determining when or if I will open it publicly. It is called timeshareexchange.us.

My Own Facebook-like Website

I wanted to have my own "private" Facebook-like equivalent just for my family, where I could store documents, photos, and videos. So, I built such a website, and its participants can rate each other's postings so that the most popular ones appear on the main page. I wanted a way to store all my favorite memories in one place. This website is not presently open to the public.

Consciousness for AI

A few years ago, I recognized that an essential step in developing Artificial Intelligence would be the ability to function with a semblance of human consciousness.[25] Incorporating consciousness into A.I. requires understanding how the human mind works, which remains an unsolved problem in I.T.

I have been thinking about this problem for years but have worked actively since early 2020. I have developed a comprehensive model of the human thought process and am currently documenting this model for publication. The book, Artificial Consciousness, was published in July 2023 and is available through Amazon.

Other Ideas

I have some ideas that I still need to pursue. One is an idea for a game. Another is a new learning system, like Neural Nets, but different. I am still determining if or when I might return to these ideas.

What I Learned in Web Development

I learned web development by taking online classes and practicing a lot. What I learned includes:

1. How to write websites – I learned HTML and CSS to create pages like you see when you call up a website in your browser. HTML is what a user sees. CSS is shorthand for formatting the pages you view that can be carried over from page to page, such as colors, lines, fonts, headings, footers, and more. In my earlier context, this is the Input and Output. I also learned responsive coding, where the website adjusts to the screen size of the device calling the site.
2. Server Programming – I learned the PHP language, which does the processing between screen displays. This portion of the website is the "inside of the Black Box" that I described earlier. A browser page cannot see what is happening inside the server, but the server can display a web page.
3. My programming skills expanded as I wrote over a million lines of code spanning ten years.
4. JavaScript – a language that communicates between the browser and the server. A software engineer, Brendan Eich created it while working at Netscape Communications Corporation in the mid-1990s. He must have worked for one of those risk-takers I mentioned earlier. This language allows the programmer to pull data from the server and display it on the displayed page rather than showing a new page. It also allows the HTML page to modify itself. I use it to suppress or add specific input fields on the webpage without accessing the server.

5. How to build a self-service website – integrated with credit card processing without storing any credit cards or related personal data.
6. How to encrypt code that contains proprietary algorithms on the server itself and have it run in a way that disguises the code from hackers even if they get into the server.
7. How to read, store, and display different forms of media, text, photos, and videos.
8. How to generate a Word Document from a PHP program (see Job Descriptions in Job Leveling)
9. I'm still learning.

CONCLUSION

One part of my experience was living through increasingly elaborate iterations of the same applications under different technologies and programming languages. For example, Siebel was written in Visual Basic. Sales Force would later be rewritten as a web-based application (HTML/CSS, Python, phone app, and others.) The evolution of programming languages has seen a progression across multiple phases. As exemplified by my experience at Stat Tab, we relied on machine languages in its early days. These later gave way to compiled languages such as COBOL and FORTRAN.

Modern languages like Python and PHP represent the latest phase of this evolution. Unlike their predecessors, these languages can run on multiple computer brands and consolidate related machine code into 'functions' or 'subroutines.' Any programmer can invoke these capabilities as needed, enhancing efficiency and flexibility.

While this evolution has significantly advanced technology, programmers are still coding in COBOL and FORTRAN even today. Some federal government systems are written in COBOL. Which language is used is not as important as whether there is a compiler to convert it to machine language. I always believed that a compiler could be created to write machine code from English words, like FOCUS came close to doing. Now AI is writing code from English instructions.

My Languages

Along the way, I have written in the following digital languages:

1. Fortran I, II, III, IV, V (the latest is Fortran 18),
2. Basic,
3. Autocoder,
4. BAL (Basic Assembler Language) – low-level IBM 360 language,
5. SAS,
6. Cobol,
7. RPG I, II, III,
8. MATLAB,
9. Oracle,
10. C/C+/C#,
11. Visual Basic,
12. ACP,
13. ASP.net,
14. JavaScript,
15. Html5/CSS,
16. PHP,
17. JavaScript,
18. SQL/MySQL4

Last year, I signed up for a beta test of IBM's Osprey Quantum Computer and wrote a couple of programs on it to gain an understanding of Quantum Computing.[18] While it has a way to go before it becomes mainstream, its potential is staggering. What interested me was that programming a quantum computer harked back to the analog computing class I took in 1963 at IIT. It uses some of the same analog techniques. As I have often said about the technology business, "What goes around comes around."

I have yet to delve much into structured or object-oriented languages, like Pascal, LISP, or other older languages like ALGOL, SNOBOL, and PL/I. But I have taken a cursory look at the current languages like Python and RUBY. I may learn Python next.

Programming languages come in four types: Functional, Imperative, Object-Oriented, and Procedural. The first two types are great for tasks that must be split up and run simultaneously on different computer processors. This process is like having multiple cooks in a kitchen, each working on another part of a recipe but all contributing to the same dish.

I usually write in the Procedural style. This programming choice is like following a recipe step-by-step. However, I find Object-Oriented (OO) coding tricky. OO coding is like having different recipes that can share ingredients and steps. My difficulty with OO coding comes from being dyslexic and the unique way I've learned to understand how computers work over the years. While I can explain how OO coding works, like describing a recipe to someone, I'd have difficulty writing a program in it, like cooking the dish myself.

I have written operating systems, compilers, drivers, and graphic interfaces. I have written on all levels of programming.

One aspect of my coding that distinguishes me from other programmers is that I "generalized" my code. I would store all fixed values that set limits on the processing in separate variables. This method simplifies expanding the program's limits by changing the variables' values. For example, if a simulation were set to run for 1,000 iterations, I would set a variable, like $slim=1000, and use $slim wherever I would have used the 1,000. Then, the entire code section would be modified by simply changing that variable. This example is a simplified one. In actual practice, I generalized all my code so that adjustments to the code could be easily made.

Programming languages today operate at different levels, from Machine Language, which is the actual instructions that a computer executes, to C/C+, which is medium level, having built-in functions (groups of canned code), to high-level languages like Python and PHP. If you begin by understanding machine language, which is the most basic form of programming, it can provide a solid foundation for grasping more complex languages. Writing code in machine language requires a lot of effort and lines of code, which helps you appreciate how higher-level languages can simplify your work. I often joke that programmers are "inherently lazy" - always seeking the most efficient solutions, the path of least resistance. That is what makes them good programmers. This characteristic is precisely what higher-level languages offer - a more efficient and easier way to code.

The lowest levels typically do not need a compiler or interpreter to run on a computer and will run as fast as possible. Where processing speed is critical, write in machine language. All the other languages need more computer processing power and time to translate them into machine language, but coding goes faster. And computer processors are so fast nowadays that the need for tight efficiency is diminished compared to the past.

What I accomplished with Computers

I led the creation of four global online real-time computer systems:

1. United Airlines' Apollo System – still in use after 50 years.
2. United Airlines' FAMIS system – shut down in 2021 or transferred to outsource suppliers after United sold off its remaining kitchens to Gate Gourmet Denver (DEN) Honolulu (HNL) Newark (EWR) Sky Café Cleveland (CLE) Newrest Houston (IAH).[19]
3. RCI's Exchange System – still in use after 43 years
4. Disney Vacation Club's Member Services System – still in use after 30 years
5. I helped build and grow major-brand companies, like United Airlines, RCI, Disney, and numerous timeshare companies utilizing the strengths of computers to improve processes, support marketing and sales, and improve operating performance.
6. I have written websites for myself as my artistic expression.

What I Learned Overall

My view of computers and programming is that they are tools for communicating. They communicate with each other. They communicate with their users. Inputs are users communicating with a program. Outputs are programs communicating with Users. All that remains between these inputs and outputs are Algebra and Text Manipulation. Things are added-up, the text is parsed, and so on. Today inputting, processing, and outputting have each become separate skill specialties:

Inputting: User Interface (UI) Design, the design around inputting, is a profession today.

Processing: Server programming (PHP, PYTHON) is processing and is a profession today.

Outputting: Web Development (e.g., HTML) is a profession today.

In the early days, one programmer designed and programmed all three. Today AI is taking all this to a new level where the Processing step has been automated to eliminate programmers. But the communicating will remain.

I also view computing as a "Blackbox" process. With computing, you put something into a Blackbox, the box does something to what you put in, and you get something back from the box. The tricks or keys to inventing and designing new computer systems, programs, functions, or applications are imagining what you might get back and then working this process backward to create what happens in the box. Starting with the goal (output) in mind, you then figure out how to do that inside the box, and as you work through that, you will discover what you need to put into the box. Along the way, you will also identify what data needs to be stored and how to store it. So, it is best to start at the end and work backward.

In developing any computer system or program, remember all the supporting activity surrounding it, such as User approvals of the initial objectives, user input on the design of inputs and outputs, Use Cases, coding, testing, procedures surrounding the use once it is working, training of the users on the procedures, and more. On a diagram that I have (too detailed to include here) of this process, almost 40 steps are shown – only one is the actual coding. Use cases are a newer way of describing functional requirements, where users and systems analysts work together to identify and define all the ways a set of programs might be used. The cases then form the input to the programmers. Use Cases may be abbreviated versions of the functional specifications I described earlier.

As I mentioned earlier, there is a well-known theorem in data processing that I was the first to formalize, or at least I independently discovered it. Management always wants your project done on time, within budget, and within scope. Unfortunately, only two of these constraints can be fixed, and the other is a consequence of the first two. Management can fix any two but only two. I often depicted this theorem by drawing a triangle on a flip chart with each constraint at one of the vertices.

I also have a hat that says "Next Phase" that I used to wear to User meetings when they wanted to add enhancements or changes to a project already in progress. I would tell them that there are three kinds of modifications that we would consider once a project has started:

> Type 1 changes "break" the original commitment for the deadline and budget because they expand the scope. This change usually means reevaluating the budget.

> Type 2 changes that would be helpful to include before cutover and might be allowed in if work on that part of the project had not started yet and the work required would not substantially be increased (approximately the same breadth of scope).

> Type 3 would be noted and deferred to the next phase, assuming there will be one, or treated as a separate post-cutover modification.

Managing Programmers

I began my career as a programmer and still identify with those with that title and do that work. I hired many programmers and programming managers in my IT departments. Many companies treat programmers as weird, nerdy, or antisocial, which is rarely valid. I always considered them essential to the success of the company where I hired them. I took extraordinary measures to take care of them.

When I interviewed a candidate for a programming position, I often employed a test of their programming skills in the programming language we were using. They also usually had to have programming experience. I hated that that requirement had a built-in contradiction. How can programmers get their first programming jobs when two years of experience are required? I always remembered the manager who hired me for my first job in technology at Stat Tab. As a result of my angst about this contradiction, I occasionally bent my rule and hired or promoted a bright but inexperienced candidate into a programming or computer operations role, as I described earlier.

Today, someone wanting to break into a programming career might work at home in the evenings writing code and then be able to show what they produced. Of course, when I started, computers were large and expensive, and no one had one at home. Now it is possible to study Computer Science courses online in your spare time.

And there is much to learn about programming that cannot be found in textbooks. When I started, the best chance was to find a job where programmers were needed so badly that the company would send you to programming classes upon hire or, as in my case, start with what I knew, Fortran, and then learn on the job by reading IBM or Honeywell manuals.

When it came to promoting or hiring programming managers, I would never hire a manager who had not previously programmed for a living because, without that experience, they would not understand what programmers needed, how to evaluate their work, how to test their programs thoroughly, and how to help them in a pinch.

At RCI, I hired a handicapped programmer from a rehabilitation training program. He had a job where his working legs were required until, at age 45, he was paralyzed in a car accident. He was married with two kids. He had to learn a new profession and chose to learn programming. I could tell he was well-trained because he aced our coding tests. I paid him our standard programmer wage, which he welcomed, having been out of work for 18 months.

As a result, he regained the independence he had lost. But then the rehabilitation program that retrained him repossessed the specially configured van they had provided him. This decision left him unable to travel from home to work and back. I felt this was unfair, so I went to our boss, Jon DeHaan, and asked him if the company would help the programmer to buy a replacement van. He agreed to provide the downpayment, which was all that was needed. Only a family-run business can be that generous. The investment paid off as he became one of our best programmers.

Another time I had just hired a young programmer, Larry S., who had been married less than a year and had a newborn baby. The week he started, he was diagnosed with lung cancer, a death sentence during the 1980s. I again went to Jon and asked if I could keep him on my payroll and allow him as much time off as needed to fight his cancer. Similarly, Jon approved my request. Over the next six months, without the pressure of earning a living or worrying about money, the young man recovered and thanked us by working for RCI for years.

Another problem in building and maintaining teams of programmers, at least through much of my career, which I learned from my experience as a programmer, is the problem of wages versus skills.

As I mentioned above, when programmers first begin their programming job, they've had minimal training and little to no experience. After getting six to twenty-four months of experience, their value in the marketplace jumped as experienced programmers have been in high demand most of my life. If they received only a standard annual cost-of-living increase each year, they would be lured away to work somewhere else within two years. Programmers are motivated by pay, promotions, and training. They particularly value training because it keeps their skills current and increases their value in the marketplace.

By providing continuous skills training, we could retain and offer our employees higher pay within our company because they were more productive with training. This progressive advancement system allowed employees to move up through positions, from programmer to senior systems analyst and beyond, each role offering a higher pay scale. This strategy resulted not just in pay and title increases but also in enhanced job satisfaction and loyalty. Yes, this led to higher annual pay raises compared to non-programming jobs, but it also improved efficiency, reduced turnover costs, and maintained our company's specialized knowledge. This approach built the strongest teams and was a factor in my successes over the years.

Postscript

The loss of experience happening in America as the Boomers and Gen-Xers retire must be a drag on the future of our country. Yes, all those coming up in the ranks behind them are bringing up new skills, but will those be sufficient, and what will be lost in the process?

I am most struck by how much everything has changed over my lifetime. Here are some examples of the evolution in my lifetime:

1. User Interface. In the Early days, there were pens and paper and then typewriters and later computing machines (like adding machines but with more functions). First, we converted punched card data into lines on Green Bar computer paper via electro-mechanical machines. We transitioned the human interface to the keyboard and Cathode Ray Tube (CRT), affectionately known as "Dumb Terminals." Subsequently, our interface moved to desktop computers, and then portable devices like laptops emerged. I know of a device you wear on your fingers that will input into a computer using taps on your leg or a tabletop. Mobile phones have moved to voice control, and now laptops and TVs. Will telepathy be next? I miss personal handwritten letters.

2. Phones. My first phone was a brown wooden box mounted on the wall. It had a crank handle, a black Bakelite earpiece that I held to my ear, and a separate Bakelite mouthpiece. (Bakelite was an early plastic) It had a party line with three families on it. If one of the others was using the line, you had to wait for them to finish their call. Sometimes others would

listen in on our calls. Operators were available to whom you could talk and ask them to connect you with someone on the phone. They could find and dial a number for you. We had printed phone books that listed everyone's phone number and address. Long-distance calls were charged by the minute, and long-distance started at the next town over. Then, some phones were moved off the wall onto a table with a wire that allowed the phone to be moved around. Using landlines, we went from black phones with dials phones in color to Princess phones with push buttons. Then we got call recorders (leave a message). Then came cell phones and phone APPs. Then earbuds where people seem to be talking to themselves. Then there's Alexis and Siri. Where will we go next? Implants?

3. Television. We went from small TVs with black and white picture tubes and vacuum tubes inside, in my case operated with a large magnifying lens in front of it, to small and then large color TVs with vacuum tubes and then to TVs built solely with transistors, to large projection screens, and then thin flat-screen TVs with LCDs. Where will we go next? Individual transmissions projecting directly into our eyes or built into lens replacements?

4. Cars. Our first car was a "turtleback-shaped" Ford that my dad shared with a neighbor, and together they painted it with polka dots. Then he bought an old Model T Ford Truck. All cars used leaded gas. It had a crank to start it, which could snap back and break your thumb if you wrapped your thumb around the handle. When I was a teen, we did not have seat belts, power steering, automatic transmissions, or hydraulic brakes. Gasoline cost

only $0.25 per gallon back then. Over time, cars started adding those features, transitioning to unleaded gas and airbags. Now, we have self-driving or nearly self-driving cars with Wi-Fi and self-parking capabilities. Soon, all cars will be hybrids or fully electric.

5. Creature Comforts. We went from using fireplaces and blankets for heat to coal furnaces to gas and electric, and from having no Air Conditioning in houses or cars to having it in most homes and cars. We have disposable everything (razors, bottles, plates, towels, and more). In our vacation cabin, we used to rely on kerosene lamps for lighting. Now, we have light bulbs that can last 10,000 hours.

6. Personal Security. When I was young, I went wherever I wanted and did whatever I wanted from age eight, as long as I was home for dinner. We had guns, but not because we thought someone else might shoot us or we might need to shoot someone else. We did not lock our doors, and we trusted our neighbors. If anyone knew your parents, they had the authority to discipline you. Nowadays, we live behind walls and utilize perimeter recording devices like NEST or RING cameras, doorbells, and body and car cameras to protect ourselves. No one lets their kids out of sight; some even carry guns for self-defense. God forbid they ever mistakenly use them.

7. Health. No one had associated lifestyle with longevity. Smoking and drinking were de rigueur. Drugs were barely known. Exercise had not yet been invented as a pastime. However, the milk industry convinced everyone that milk was a necessary dietary component. Milk was deemed healthy and

was delivered by milkmen who left bottles twice a week on our doorstep, first in horse-drawn wagons, then later by trucks. I always thought it neat that the horse knew which house to stop at. Cataracts can be fixed with lens replacements and soon eyedrops. We have vaccines for many more diseases – I suffered through pneumonia, the Mumps, Chickenpox, and Measles, and lived through the era of Polio, and gladly would have preferred having vaccinations had they existed. That there are still people who have no health insurance and can be financially ruined and even made homeless if they get sick is immoral in these times. Our medical infrastructure is obsolete, too heavily dependent on drugs, and not sufficiently on natural cures, such as fasting and exercise. Hopefully, AI will soon make medical diagnoses and consider natural treatments before chemicals.

8. The computer industry has become so large, so complex, and has so many automated functions that job opportunities are both vast and limited. There are many choices of where to contribute, but your contribution will be less impactful on an individual level. Computer Applications are enormous in scope and often require hundreds or even thousands of people. As a result, the opportunities for individual contributions are shrinking and nearly gone.

There are many more examples of how change is unstoppable. Think about all the changes in one lifetime, then imagine what will happen in the next 100 years. If you want to keep up, be a change-maker, not a victim of change.

Epilog

My Advice to Young People Seeking Jobs: "Hook your wagon to a Star!" Get in on the ground floor of a burgeoning business in an industry with vast growth ahead of it. Jobs are much more fun when you are working on a "start-up," and I think you will learn the most in that environment. If you don't have the vision, ask others for suggestions, until you hear some continuity in the advice. A target will become clear. Examples might include Super Computers, Space Programs, renewable organic farming, or renewable power sources. Consider building or repairing robots as a profession. At the same time, remain human.

Be kind to others. You never know when you may meet them again. Be an ambassador for your industry. Learn to speak publicly; it elevates you above those who fear getting up in front of a crowd. If you need to learn how or are fearful; start by attending Toastmasters meetings. Find a class on "How to win friends and influence people." Get involved in industry associations. Never fear introducing yourself to someone new. They will appreciate it, and you may gain a friend. Meet as many people as possible, help them when you can, and maintain your relationships over the years. Do not drink alcohol at conferences. I have seen careers ruined by that.

Think positive. See the positive in what you do. For example, many people thought timeshares were terrible. I saw pressing people to buy timeshares as being good for them. Owning a timeshare forced the owners to take vacations with their families and did not cost all that much over time, not as much as many spend on bad habits – drinking, smoking, playing golf, and other vices.

Lastly, when you are facing something difficult or that you do not want to do, change your focus to concentrate on how relieved you will feel once you have done it. Then do it. A book called Bird by Bird for writers explains how to do a large writing project about birds, where the advice was "Bird by Bird." Apply this advice anytime you have a large project and are unsure where to start—bird by bird.

Call me a Nerd

If you see me as only a Nerd, which I admit to, please know that I have also been a Jock sometimes. I have been physically active throughout my life. Here is a list of my physical endeavors and the time spans over which I did these activities:

- In grade school, I took up archery at an adult club and got my first flyfishing rod. – I fly-fished off and on most of my life in places from Wisconsin to Wyoming to Florida. I fly-tied my first year.
- In grade school, I competed in the Soap Box Derby, finding a sponsor in the local VFW, and building my car myself.
- High school Football, played End on offense and defense, punted, kicked off, and received punts and kickoffs – 2 years.
- High school Baseball – 2 years as manager.
- High school Basketball – 1 year.
- Hunter of pheasants, rabbits, squirrels, opossums, raccoons, ducks, pigeons, and field mice – off and on from age 12 through age 16. One deer as an adult, which I regretted.
- Wrestling (high school) – 1 year.
- Gymnastics (college class) – 1 semester.

- Archery (college class) – 1 semester.
- Wrestling (college class – Gold Metal in my weight class).
- Intramural Football (College) – two years.
- Swam weakly at age 7, Learned to swim well at age 38, and swam my way out after being caught in a riptide at age 74 (another near-death experience).
- Scuba diving – certified instructor, master instructor – 3 years (2 near-death experiences).
- Jogging – 2 years (best: 6.5-minute miles x 3.5 miles).
- Weightlifting with a trainer – 3 years.
- Tae Kwon Do 1st degree (IN), 2nd degree (twice, IN & FL), 3rd degree (FL), and 4th degree (FL), master instructor, national referee, Pan Am games organizer– 16 years.
- Class in competitive fencing by a Disney Instructor.
- Ocean fishing for cod, flyfishing for sea trout, red snapper, snook, and other species. Fresh-water flyfishing for bluegill, sunfish, bass, crappy, walleye, northern pike, and others, and river fishing for 50-pound carp and catfish.
- Basketball (after work) 2x/week– 4 years.
- Mountain Climbing -- Four attempts with three summits: Mt. Rainier (Washington state), Mt. Washington (New Hampshire), Mt. Santis (Switzerland), and the highest Mt Kilimanjaro 19,532 feet in Tanzania), – over three years.
- Tennis – 3 times/week at 2.5 hours each – 4 years to age 73.5.
- (now) Walking 20,000-70,000 steps (35 miles) per week.
- (now) Weight-lifting – once or twice per week (moving 20,000 to 30,000 pounds per workout).

Other notable activities:

- I skipped kindergarten and started first grade a year too young.
- Wore hand-me-down clothing.
- Played son of Medea – commercial play in Pittsburg at age seven.
- Learned to drive a workhorse plowing a field (Gee, Haw!) at age twelve.
- I collected stamps, which taught me geography, names of countries, foreign currencies, and the word philately. Also, I collected coins.
- Worked as a paperboy which taught me how to run a business.
- Milked cows by machine and by hand.
- Drove a tractor disking a 40-acre field at age 11.
- I once walked behind a tractor and plow, picking up large rocks with another teen my age for seven hours.
- I mixed gunpowder for explosives, helped fire off a city's July 4th fireworks, and mixed rocket fuel once I launched a six-foot, 4" diameter rocket that traveled over a mile, horizontally, unintentionally.
- I played a lead role in two H.S. plays - one a musical, Brigadoon, where I played three parts: a young man, a dancer, and an old man, and the other the Madwoman of Chaillot.
- I was once the rear-end of a horse in a variety show rehearsal.
- Learned how to cut gems.
- Learned how to fix radios and TVs.
- I never did drugs of any kind. I drank too much alcohol. I smoked my first pack of Lucky Strike cigarettes at age eight and smoked after that too much until age 32, when I quit both. I wish I had done neither, ever.

- I raised two kids – with my wife, plus four Labrador retrievers (two rescues) and one white fluffy mutt (also a rescue).
- Trained to become a Certified Gemologist -- Traveled to Colombia twice to buy emeralds) and once around the world to Sri Lanka to purchase rubies.
- Traveled to every state in the union except North Dakota and Alaska, plus internationally: Canada, Mexico, Aruba, Bahamas, Venezuela, Colombia (2x), England, France, Germany, Denmark, Belgium, Sweden, Spain, Italy, Switzerland, South Africa, Kenya, Tanzania (2x), India, Sri Lanka, Singapore, and Japan.
- Fraternity - (college) Member two years, Board Chairman Phi Kappa Sigma two years.
- Beer, Wine, Scotch drinker – 15 years, quit at age 30.
- Organizer of beer surveys (2 years).
- Winemaker (3 years).
- Played bass in a wedding band (2 years).
- Played classical guitar (10 years).
- Smoker – Cigarettes, Pipes. Cigars – 10 years.
- Organic Gardener (off and on over many years)
- Woodworker (3 years).
- Writer (10+ years) – wrote and published eight books, so far.
- Two more near-death happenings – two pulmonary embolisms after arthroscopic knee surgery (25% mortality rate) and a severe intestinal infection in May 2022, where I truly felt myself sinking into death but pulled myself back by sheer willpower.
- I am a proud father of a son and daughter, a grandfather of five, and a great-grandfather of two.

Final Notes

While it should be evident, all the above could not have been accomplished without the love and support of my beautiful wife of 63 years and still going. She has been my guardian angel, my head cheerleader, my best friend, my partner, and overall, my best girl. She always let me take my head as a successful rancher might do with a Mustang. And I love her for that, more and all.

For more information on learning to lead, see my book, *Extreme Leadership*, available on Amazon.[25] You can see all my books on www.charlespattonbooks.com.

Sources

These are references to websites or articles used to confirm or source specific details.

Web Links

1. Computer History Museum. (n.d.). *1960 Timeline.* Retrieved from www.computerhistory.org/timeline/1960
2. Next Gen Personal Finance. (n.d.). *What percent of college graduates end up working in their field of major?* Retrieved from www.ngpf.org/blog/question-of-the-day/qod-what-percent-of-college-graduates-end-up-working-in-the-field-of-their-major/
3. Call Centre Helper. (n.d.). *Aspect Telecommunications: The history of the call centre.* Retrieved from www.callcentrehelper.com/the-history-of-the-call-centre-15085.htm
4. Wikipedia. (n.d.). *History of programming languages.* Retrieved from en.wikipedia.org/wiki/History_of_programming_languages
5. LinkedIn. (n.d.). *J. Sacra - Experience details.* Retrieved from www.linkedin.com/in/jsacra/details/experience/
6. National Women's History Museum. (n.d.). *Hedy Lamarr Biography.* Retrieved from www.womenshistory.org/education-resources/biographies/hedy-lamarr
7. New Relic. (n.d.). *Python programming styles.* Retrieved from newrelic.com/blog/nerd-life/python-programming-styles
8. Federal Register. (2016, August 3). *Special conditions: Embraer S.A. Model EMB-545 and*

EMB-550 airplanes, synthetic vision system and flight display. Retrieved from www.federalregister.gov/documents/2016/08/03/201 6-18447/special-conditions-embraer-sa-model-emb-545-and-emb-550-airplanes-synthetic-vision-system-and

9. Resort Trades. (n.d.). *Pioneer Series: Edwin H. "Ed" McMullen Sr., RRP.* Retrieved from resorttrades.com/resort-trades-pioneer-seriew-edwin-h-ed-mcmullen-sr-rrp/

10. Hospitality Net. (n.d.). Retrieved from www.hospitalitynet.org/editorial/4012295.html

11. New England Equipment Distributors. (n.d.). *Equipment details.* Retrieved from www.neequipdist.com/equipment/details.cfm?eid=7 49

12. *(Omitted)*

13. *(Omitted)*

14. University of Oxford. (n.d.). *Brief history of dyslexia.* Retrieved from dyslexiahistory.web.ox.ac.uk/brief-history-dyslexia

15. Wikipedia. (n.d.). *Edward Carlson biography.* Retrieved from en.wikipedia.org/wiki/Edward_Carlson

16. ScoutSmarts. (n.d.). *12 Scout Law Principles.* Retrieved from scoutsmarts.com/12-scout-law-principles/

17. *(Omitted)*

18. IBM Newsroom. (2022, November 9). *IBM unveils 400+ qubit quantum processor.* Retrieved from newsroom.ibm.com/2022-11-09-IBM-Unveils-400-Qubit-Plus-Quantum-Processor-and-Next-Generation-IBM-Quantum-System-Two

19. Live and Let's Fly. (n.d.). *United Airlines closing flight kitchens.* Retrieved from liveandletsfly.com/united-airlines-closing-flight-kitchens/

20. Computer History Museum. (n.d.). *1958 Timeline.* Retrieved from www.computerhistory.org/timeline/1958/

21. Wikipedia. (n.d.). *ILLIAC IV.* Retrieved from en.wikipedia.org/wiki/ILLIAC_IV

22. Wikipedia. (n.d.). *Electronic Data Systems.* Retrieved from en.wikipedia.org/wiki/Electronic_Data_Systems

23. IEEE. (n.d.). *History of Wi-Fi infographic.* Retrieved from www.ieee.org/content/dam/ieee-org/ieee/web/org/about/wifi_history_infographic.pdf

24. BBC News. (n.d.). *Technology article.* Retrieved from www.bbc.co.uk/news/technology-65401783

25. Amazon. (n.d.). Extreme Leadership: Leaders achieve results. Retrieved from www.amazon.com/Extreme-Leadership-Leaders-achieve-results/dp/B0BFTWJBXZ

Other References

1. Platt, Charles, "Back to the Analog Future", Wired, May 2023, from CNN News post of May 1, 2023

www.ingramcontent.com/pod-product-compliance
Lightning Source LLC
Chambersburg PA
CBHW071321210326
41597CB00015B/1303